*Process Engineering
for a Small Planet*

# Process Engineering for a Small Planet

## How to Reuse, Re-Purpose, and Retrofit Existing Process Equipment

**Norman P. Lieberman**

**WILEY**

A JOHN WILEY & SONS, INC., PUBLICATION

*Library of Congress Cataloging-in-Publication Data:*

Lieberman, Norman P.
  Process engineering for a small planet : how to reuse, re-purpose, and retrofit existing
process equipment / Norman P. Lieberman.
       p.   cm.
  Includes bibliographical references and index.
  ISBN 978-0-470-58794-2 (cloth)
  1. Chemical plants–Equipment and supplies.   2. Chemical plants–Remodeling for other use.
3. Salvage (Waste, etc.)   I. Title.
  TP157.L474   2010
  660′.283–dc22                                                      2010001896

Printed in Singapore

10  9  8  7  6  5  4  3  2  1

*To the memory of shift foreman Mohamed Lee, born 2012–died 2090, who shouted: "Shamil, you bloody fool! Throttle on the inlet guide vanes to the blower; don't open the discharge atmospheric vent. Didn't you read Lieberman's book? You idiot! You're wasting amps!"*

*To my daughter, Irene, who assembled and reassembled this manuscript.*

*To my wife, Liz, who inspired this project: "Norman! Stop whining about the environment and do something. Maybe write a book."*

*To my mother: "Your cousin's coat is just like new. We're poor people. We have to get by with what we've got. You'll wear the coat and like it. And your aunt sent it especially for you. Look, even all the buttons still match."*

# *Contents*

# *Foreword*

Mr. Lieberman's books, his knowledge and understanding of, and expertise in, the field are second to none. His writing is one of the best (if not the best) in the industry, unsurpassed for clarity, and at the same time absorbing and entertaining. You do not doze off reading his books. His hands-on approach makes his writing understandable by all and suitable for all levels, from the director and technical expert all the way to the novice and the nondegreed operator. People who have been in the game for decades will pick up his books and be able to gain new insights. His practical approach will make it possible for every reader to find something that she or he can use and apply immediately in their own fields.

This book represents a slight departure from Lieberman's normal writing. His writing usually teaches the practical art of chemical engineering and effectively passes on his vast experience to readers, preaching understanding and excellence in engineering. In this book, his teaching is supplemented with a message leading a crusade to "save Planet Earth from destruction, global warming, and environmental hazards." The beauty of this book is that he shows that following *good* chemical engineering practices paves a path to protecting the Earth. In contrast, he demonstrates how poor engineering and wasteful practices are one of the major hazards to our planet. He gets across the message: Protecting the Earth is not in "their" hands; it is not what "they" need to do. It is our responsibility as engineers. It is our duty to identify bad and wasteful engineering practices and stand up to them. Do not use the excuse that a manager above you set the tone. Do not hesitate to stand up to her or him. As a professional, fight such a person with good technical work. With his usual practical writing style and Lieberman humor, the author gets this powerful message across very clearly. Better than many of the politicians who talk about it, he is *doing* something about it. The political solutions that are promoted in the media often lack a good engineering basis. The need is for engineering solutions, not political ones, and it is we who must come up with them.

There is no question that this book will be a great addition to the literature and is much needed by the industry. Unlike his other books, which are used primarily by industry, this book may make some inroads into academia. The "save the Earth" message will make it fashionable and popular, and academics frequently welcome new ideas, such as engineering rather than political solutions to our global warming problems. Lieberman's usual audience of operation engineers, control engineers, process engineers, design engineers, process operators, and research engineers will be very interested—especially once the enormous costs and undertakings of carbon-capture technology are appreciated—and people will be seeking alternative cheaper solutions, such as those the author advocates. His message is universal and I believe will be supported almost unanimously by the process engineering community.

HENRY KISTER

*FLUOR*

# *Preface*

Life goes on. And time goes on. The seasons change one into the other. The hills, the rivers, and the seas will survive, but perhaps without us.

It's a process problem. Our small planet, our home, is a rather large process plant responding to a number of process variables. One of these new process variables is man.

As I explained in my previous book, *Troubleshooting Process Plant Controls* (John Wiley, 2008), any unconstrained process variable may result in a **positive feedback loop**. Our human activities in the past 70 years have introduced just such an unconstrained variable into the dynamics of Earth's operation.

At the heart of the problem is the hydrocarbon refining and petrochemical processing industry—an industry that I, after 46 years of activity, have become identified with, owing to my design, teaching, writing, and field troubleshooting.

I can see in retrospect from the perspective of the process engineer that I have made a significant and sad contribution to the environmental positive feedback loop our little planet has entered. If our Earth is a complex *process facility*, it's clear that we can no longer escape the consequences of our actions. But how can we work to mitigate these consequences?

What's to be done? What can I and my co-workers in the hydrocarbon process industry do to retard environmental degradation? This book, *Process Engineering for a Small Planet*, focuses on this question. The text is basically technical. As in my other five books, the format includes stories drawn from my personal experiences in petroleum refineries, chemical plants, and natural gas production wells.

The lesson I teach is that we have to learn to use our existing plant facilities to expand production and improve energy efficiency rather than constructing new pumps, compressors, distillation towers, and vessels. The lesson is that we live on a small planet with limited air, water, and mineral resources. But rather than lecture about morality, I have presented a series of technical and engineering examples to best fulfill my mother's advice: "Norman, we're poor people! We can't afford anything new. We've got to get by with what we have."

It's true! We live on a small planet. We've got to get by with the plant facilities we already have. This book will suggest how we in the hydrocarbon process industry can conform to mom's instructions.

## For Whom Intended

Since 1983 I have been teaching a seminar entitled "Troubleshooting Process Operations." About 16,000 technical personnel have attended. In the last few years, I have devoted class time to detailing the growing environmental hazards—chiefly $CO_2$, methane, and $NO_x$ emissions—that we, as process technicians, engineers, and managers, are generating.

The response to my challenge is the question: What's to be done? Engineers, operators, production personnel, and management in the hydrocarbon production and processing industry do not design solar panels, windmills, or fuel-efficient cars. We do not develop hydrogen fuel cells or methods to convert algae to biodiesel.

I am asked: "In the scope of our work as process plant personnel, what can we do?"

*Process Engineering for a Small Planet* is my response. What can we in the hydrocarbon process industry do to reduce environmental degradation? Can we contribute to the solution, or are we to be swept along by the tide of events?

We are surrounded by propaganda, which is misleading and often simply lies:

- Clean coal
- The hydrogen economy
- Ethanol as green fuel

Process engineers and operators do not need to look to exotic technology to make our contribution to combating the environmental crisis. Fully 10% of fossil fuels are consumed in the production, refining, and processing of coal, crude oil, and natural gas. Huge amounts of steel, copper, and cement are consumed to construct new, and often unnecessary, process equipment.

*Process Engineering for a Small Planet* is a handbook of ideas as to how to operate and retrofit existing process facilities to:

- Save energy.
- Reduce greenhouse gas emissions.
- Expand existing plant capacity but without installing new equipment.
- Reduce corrosion and equipment failures.

The text is technical. However, as in my other books:

- *Troubleshooting Refinery Processes*, 1980, PennWell

- *Troubleshooting Natural Gas Processing: Wellhead to Transmission*, 1985, PennWell
- *Process Design for Reliable Operations*, 2nd ed., 1989, Gulf
- *A Working Guide to Process Equipment*, 3rd ed., 2007, McGraw-Hill
- *Troubleshooting Process Plant Control*, 2008, Wiley
- *Troubleshooting Process Operations*, 4th ed., 2009, PennWell

the format includes stories drawn from personal experiences in petroleum refineries, petrochemical plants, and natural gas wells. The manuscript is devoid of complex mathematics and pedantic paragraphs. Technical text that is presented in a conversational tone is unusual, but *Troubleshooting Process Operations*, written in this style, has been on PennWell's best-seller list for 28 years.

## Disclaimer

I have represented the technical facts in this book to the best of my understanding. However, references to places, names of refineries, and of individuals have been chosen at random. Any reference to people or places that corresponds to any actual people or places is purely a coincidence. However, all related experiences are technically complete and correct. Perhaps you will recognize one or more of these stories from your own experiences and think it is your story that I am writing—maybe it is and maybe it isn't. I have seen many of these scenarios in more than one place. As further evidence, in one of my seminars, an attendee came up to me on a break and asked who had given me his story to tell because the circumstances were so similar—it was not his story but it took a while to unravel the confusion. Please contact me if you have a similar question.

Engineering formulas presented here are usually approximations that have been simplified to promote comprehension at the expense of accuracy. References to more detailed texts are provided if accurate calculations are required.

The stories presented are stated in the context of my having initiated the improvements. Often, I was just a participant in implementing other people's concepts and have failed to assign appropriate credit. Such concepts have not been stolen from my clients—they have simply been borrowed and will be returned one day soon.

## Process Problem Inquiries

If you have a process engineering question related to this or any of my other books, please call me at:

- 1-504-887-7714

Or, you may fax me at:

- 1-504-456-1835

There is no charge for such consultations. However, please don't send emails. I can't type and do not plan to start learning to do so at my advanced age.

Before you phone or fax, you would be best to conduct a field survey to collect the relevant data. While you are collecting the data in preparation for our phone consultation, you may well stumble across the bit of information that you lacked to solve the problem. Then you can forget about bothering me and thus avoid destroying my sense of peace and well-being. However, if you are absolutely determined to send me an email, my address is

- norm@lieberman-eng.com

NORMAN P. LIEBERMAN

*May 2010*

# Introduction

# Turning of the Tide

Morning had broken, like the first sunrise. The hidden fjord was bright and still in the sunshine. I paddled my red kayak easily between sheer rock walls cut by the strong hand of a long vanished glacier, toward the distant snow-capped peak. Space and time had lost all meaning.

Enclosed by the towering fjord, with no possible landing beach, I reversed course and headed back to camp. Paddling hard, I glanced at a tree growing from the cliff wall. The tree hadn't moved. In 20 minutes I had not progressed 20 feet. The tide had turned against me.

The tide would turn again, but then it would be night, which would bring a cross-wind and a rip-tide. Allowing the tide to sweep me farther into the fjord was an option with unknown consequences. Trapped in the fjord at midnight, with the black waves rebounding from the granite walls, was a fatal prospect.

What was I to do? I paddled with all my strength. Alone and afraid, I prayed for divine deliverance. But the Universe is vast, and the Creator did not answer my plea right away, but left me there awhile to think.

As a species we too are trapped, with time and tide turning against us. A tide of solar energy stored as carbon has carried us into a dead-ended technology. Oxidizing fossilized carbon to provide energy for 7 billion people in our growing industrialized economy is madness. Timing is uncertain, but the outcome can be calculated.

Like my experience in the misty fjords in Alaska, we'll have to rely on our own resources and with the tools at hand. Otherwise, the tide of greenhouse gas from

*Process Engineering for a Small Planet: How to Reuse, Re-Purpose, and Retrofit Existing Process Equipment,* By Norman P. Lieberman
Copyright © 2010 John Wiley & Sons, Inc.

tar sands, shale oil, coal, peat, natural gas, methane hydrates, heavy crude oil, and nitrous oxides will sweep us into oblivion.

Some are waiting for new technology. Clean coal. $CO_2$ sequestration. Biodiesel from fermented cellulose. Some are waiting for Divine intervention. Others are hoping for offshore drilling. But I can't wait. Let's all face the truth. We must make progress with what we have at hand today.

The hydrocarbon process industry is at the center of the crisis. I just visited a coke gasification plant in Kansas. The plant converts petroleum coke to hydrogen. Great! Except that it vents 1 mol of $CO_2$ per mole of $H_2$. Suppose that this process becomes the long-term solution to the energy crisis in the United States? Tar sands, ethanol, shale oil, natural gas to liquids, fuel cells, the hydrogen economy—are all just as bad.

There are 10,000 to 20,000 senior leaders in the hydrocarbon processing industry. As far as tenure goes, I'm in the top rank: 45 years in refining, petrochemicals, and natural gas processing. Perhaps society as a whole cannot alter our carbon-dependent economy. But we, the process engineers in the hydrocarbon industry, can make a difference.

My book is not a solution. It's not even a plan, or a start, or the beginning. It's just a prayer whispered into the wind.

Let's use the process equipment that we have. Let's use our chemical engineering skills to avoid building new facilities, but operate our existing plants in an efficient manner. The expansionist economy we have created has to be reversed. How can this be done? Well, I've 45 years' worth of experience to share with you.

I paddled my kayak back for five hours and eventually won through. It was a struggle the whole way. Nor will it be easy for us to escape the mortal grip of the carbon economy that we, the process engineering community, have created. But I, for one, am going to try. This book is my contribution to that goal. If we don't try, we will surely fail. Ladies and gentlemen, the tide of time is not on our side.

# Chapter *1*

# *Expanding Fractionator and Compressor Capacity*

Last night, in my dreams, I traveled through time and space. The universe was vast: dark and still. In my dream I ascended Mount Olympus, where King Zeus, son of Cronus, Queen Hera, the Earth Mother, and Pallas Athena, Goddess of Wisdom, reign over the affairs of man and beast. Father Zeus and other immortals had gathered around a pool of crystal clean water. Peering into the pool, I could see images of my home, New Orleans, submerged beneath the waves of the Gulf of Mexico. King Zeus rippled the water with a wave of his hand. Now I could see Greenland, bare of its ice cap. Zeus waved his divine hand again and Kansas appeared. Not green with corn and soybeans, but as a desiccated windblown desert.

Athena, Goddess of Wisdom, looked sadly at me and said, "Thus have humankind's actions destroyed the creation of the Titans; the Blue Planet; the Pearl of the Universe. Look deeply into the sacred waters and learn the folly of human ways."

And as I obeyed the command of the daughter of Zeus, I saw a six-drum delayed coker in Los Angeles. Father Zeus spoke thus: "Norman," Zeus commanded, "Tell me about your life."

"It's a long story, Son of Cronus," I said.

"Not a problem," responded Zeus, "We have all eternity."

"Okay. Well, I was born in 1942 in Brooklyn. I married and had three children. I studied chemical engineering and graduated in 1964 from. . . ."

"Norman," King Zeus interrupted, "I know all that. What I'm interested in is the C-301A, the new coker fractionator you designed for the Saturn refinery in Los Angeles."

*Process Engineering for a Small Planet: How to Reuse, Re-Purpose, and Retrofit Existing Process Equipment,* By Norman P. Lieberman
Copyright © 2010 John Wiley & Sons, Inc.

"Well, this was a 26-ft-I.D. by 112-ft tangent-to-tangent tower that. . . ."

"Tangent to tangent was 112 ft and 4 in.," Hera corrected.

"Yes, Immortal Queen Hera, it was 112 ft and 4 in. C-301A was a new tower. The largest coker fractionator on Earth."

"And how about C-301, the existing 17-ft-I.D. tower?" asked Pallas Athena.

My exit interview had taken an unpleasant turn. In 45 years, I had designed hundreds of distillation towers. Why did Zeus have to select this tower—the project that I would least like to dwell upon? Especially the fate of the old C-301 coker fractionator.

"Rulers of Heaven, it was all so long ago. Anyway, it wasn't my fault. I had a contract for $132,000. Don, the project manager, told me what Saturn wanted. It was Don's fault. Not mine. The scope of work was defined by my client. I've forgotten the details. How about my revamp of the El Dorado polypropylene plant? Would you like to hear about. . . ."

"Norman," Zeus thundered, "Thou hast sinned. Man was made in God's image, the steward of the Earth. Have you been a good steward of this small planet, unique unto all the heavens?"

## SATURN'S COKER FRACTIONATOR

In 1966, I had revamped the Amoco viscous polypropylene unit at El Dorado to increase its capacity by 60%. Amoco was going to build a new plant to get the extra 60%. But I realized that I could "de-bottleneck" the unit by 60% by converting a natural-circulation refrigerant evaporator into a forced-circulation refrigerant evaporator (see Chapter 12). All I needed was a new refrigerant pump and some 6-in. piping. But Zeus wasn't interested in that project.

Actually, I remembered the coker project in Los Angeles in detail. However, my plan to blame Don, the project manager for this fiasco, was a nonstarter in the eyes of the Immortals. So here's what happened. Maybe you can say a prayer for me.

## OBJECTIVES OF DELAYED COKER EXPANSION

Figure 1-1 is a simplified sketch of a refinery delayed coker. The coker had a capacity of 60,000 bsd, as limited by the flooding in the fractionator. The objective of the expansion project was to increase the capacity to 75,000 bsd. I had been retained to prepare a process design to achieve this 25% expansion. My plan was to reuse the existing C-301 fractionator by:

- Increasing the fractionator operating pressure by 8 psig.
- Reducing the recycle of coker gas oil to the coke drum by leakproofing the gas oil pan chimney tray.
- Minimizing the use of unneeded purge steam used at various points associated with the coke drums.

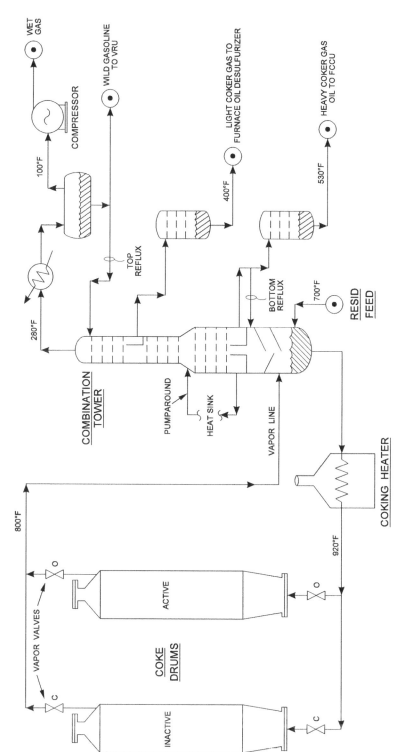

**Figure 1-1**  Simplified process flow diagram of a delayed coker.

5

- Increasing heat extraction on the gas oil pump-around loop.
- Drilling holes near the tray rings and tray panel seams in the tray panels that did not have any room for valve caps, to optimize the hole area at 13 to 15% of the tray active area.
- Sloping the tray downcomers, to increase the tray deck area.
- Reducing the outlet weir heights to a minimum on the critically loaded trays.

## CHANGING TRAY PANELS

As an alternative to modifying the existing tray panels, one could change the tray panels without modifying the existing tray rings supports and still reuse existing downcomers. When done properly, an increase of 5 to 15% in the tray vapor-handling capacity will result. Changes that are required include:

- Cutting off the bottom edge of the downcomers (about the bottom 4 in.) and restricting the downcomer bottom to preserve the downcomer seal.
- Adding a push-type valve tray panel below the downcomers.
- Replacing the tray panels with Provalves (from Koch-Glitsch) or MVG Grid Trays (from Sulzer-Nutter).

Part of the extra capacity results from using the area under the downcomer for vapor flow. Part comes from pushing the liquid across the tray deck, which equalizes the liquid level on the tray deck and thus promotes more even vapor flow to each tray.

## REDUCING THE GAS OIL CONTENT OF FEED

My other proposals would decrease Saturn's excess coker feed by reducing the gas oil content of the delayed coker's feed. The coker feed pumps used a gas oil for seal flush material. These were older pumps, with archaic mechanical seals. Four pumps were involved. Each had both an in-board and out-board mechanical seal, for a total of eight seal flush points. Each seal flush point consumed about 3 gpm of gas oil:

$$(8)(3 \text{ gpm})(1440 \text{ min/day}) \div 42 = 840 \text{ bsd}$$

(*Note:* An idle pump uses 60% of the seal flash used by a running pump.)

Thus, 1% of coker feed was recycled gas oil. I could change the older seals to modern seals that use high-pressure nitrogen as a barrier fluid. Changing the seals would be inexpensive compared to the cost needed to replace the existing pumps. (*Note:* The Eagle-Burgman seal is a good choice.)

I had also noted that the vacuum tower stripping section was not using enough stripping steam. By increasing the flow of the vacuum tower bottoms stripping steam,

I could reduce the gas oil content of the delayed coker feed from 12% to 10%. This would reduce the coker feed rate by about 1200 bsd.

Finally, the flowmeters on the coker heater pass orifice meter connections were purged with gas oil. There were eight passes, each with two orifice tap connections. Rather than continuously purging these 16 orifice tap connections, seal pots packed with gas oil could be used. This would reduce the gas oil content of coker feed a further 250 bsd. Overall, these three indirect methods would decrease the required delayed coker capacity by an additional 3 to 4%.

I had thought that combining all these modest changes would increase the coker capacity by 20 to 25%, or about 72,000 to 75,000 bsd. I knew that raising the fractionator pressure, which would also increase the fractionator capacity, would not be an acceptable option if it also raised the coke drum pressure. The problem was that each increase of 8 psi in coke drum pressure would also reduce the delayed coker liquid yields by about 1.5 liquid volume percent. However, I had also observed that the current differential pressure between the coke drums and the fractionator was about 12 psi. Most of this $\Delta P$ (see Figure 1-1) was due to not having full ports in the coke drum overhead vapor valves. The valve port sizes were only 70% of the line sizes. This means that the flow area through the orifice was only half of the flow area through the process lines. As $\Delta P$ varies with

$$\Delta P \text{ proportional to velocity squared}$$

I could eliminate the majority of the pressure loss through the coke drum vapor lines by replacing the existing vapor valves with full ported valves. Thus, the coke drum pressure would barely change, even though the fractionator pressure would increase from 20 psig (35 psia) to 28 psig (43 psia).

Tower capacity varies inversely with the square root of the absolute pressure. Thus, my single idea of increasing the fractionator pressure by 8 psi would increase the tower's capacity by 11%:

$$\sqrt{43 \div 35} = 1.11 \quad (\text{i.e., } 111\%)$$

## "JUST DESSERTS"?

Don, the Saturn project manager, obtained a cost estimate of $8 million for my design. But my design was rejected by Saturn for several reasons:

- I could not provide an absolute guarantee that the existing coker fractionator would not flood at 75,000 bsd of feed.
- The Saturn project planning department had budgeted $100 million for this project and capital investment allocations could not be transferred to next year.
- The Saturn plant manager, Larry Overbourne, wanted a new tower.

"It's okay, Norm," Don explained. "We realize that a new contract is required. I've already generated a new purchase order for your additional work. Just design the new coker fractionator so that it won't flood at 80,000 bsd and 20 psig operating pressure. The bigger the better. That's the way Mr. Overbourne thinks."

"But, Don," I responded, "I've spent so much time on the revamp of C-301. I think it will do the job. The capacity of the unit will be limited by the size of the coke drums to less than 75,000 bsd anyway. Mr. Overbourne's 80,000 bsd feed rate target is completely unrealistic. The coke drums will limit unit capacity to. . . ."

"Norm, the new purchase order for your work is $132,000—a lump sum," Don said. I couldn't think what to say. That's a lot of money and I knew I could do the entire design in just two weeks. So I changed the subject.

"Look, Don, the wet gas compressor will not be big enough. Not with the fractionator running at only 20 psig and 80,000 bsd of feed. That's my main reason for raising the coker fractionator pressure by 8 psig. The resulting higher wet gas compressor suction pressure, from 10 psig to 16 psig, will allow me to raise the unit charge from 60,000 bsd to maybe close to 75,000 bsd. Also, I could. . . ."

"No, Norm. You're not listening," Don interjected. "Mr. Overbourne also wants a new 12,000-hp compressor."

"But, Don, there's nothing wrong with the existing 9000-hp compressor. Anyway, the electrical substation won't handle the extra load."

"Lieberman," Don concluded in a firm voice, "I've a meeting to go to. So let's wrap this up. Listen to me:

- *First point.* The $100 million includes the cost of all electrical work, especially a new substation.
- *Second point.* If you don't want the work, Wild Horse Engineering will be happy to take over.
- *Third point.* You should show more respect for Saturn management."

So I signed the contract. And now I had to answer for the new C-301A fractionator. But it wasn't my fault. Maybe Moses dropped the tablet with the Eleventh Commandment: "Thou shall not waste the resources of the Earth." But that's not my fault either. It's all Don's fault. He led me into temptation.

## WET GAS COMPRESSOR

I guess it's true. The Immortals know the evil that dwells in our hearts.

"Norman," said the Son of Cronus, "did you know that 16,000 tons of iron ore had to be torn from your small planet to fabricate the new fractionator? Plus 16,000 tons of No. 9 coal. All for what purpose?"

"Well, Master of Mt. Olympus. All for no purpose. As you see, I would trade all of the $132,000 just to get my kayak to the shore. And the L.A. delayed coker was limited to 70,000 bsd of feed by the capacity of the existing coke drums, which with relatively minor process changes, the old C-301 tower could have handled."

I could see, though, that it was best not to mention again that it wasn't my fault. Not only because Zeus didn't believe me, but because in my heart I knew—and had always known—the truth.

"Zeus, forgive me. I am at fault," I admitted.

"And how about, Norman, the new K-301-A, 12,000-horsepower wet gas compressor?" inquired Athena.

"To answer that question, I'll have to refer to the second law of thermodynamics."

"Yes, the second law of thermodynamics." Hera seemed pleased. "The Titans created the laws of thermodynamics when they separated light from darkness. Yes, they created the laws of science for humankind to use and not to abuse. But you have perverted science in sinful ways."

I had told Don that the capacity of the coke drums was inconsistent with a new and much larger, 12,000-hp centrifugal compressor. It would not really matter, I explained, if the compressor were oversized if we used a variable-speed driver. There were three variable-speed options:

- *Steam turbine.* We could install a 400-psig motive steam turbine, exhausting to a surface condenser operating under vacuum conditions. But Don said that the refinery was short of cooling water and the proposed surface condenser would consume 8000 gpm of cooling water.

- *Gas-fired turbine.* There was plenty of coker fuel gas to burn. But this would require a permit, as a new emission source, from the state of California, which could take years.

- *Variable-speed motor drive.* This would involve the purchase of a relatively expensive motor, as the frequency of electric power to the motor would have to be varied. Unfortunately, Larry Overbourne, the refinery manager, did not like variable-frequency speed control of large motors in critical services.

So we would have to use an ordinary 15,000-hp motor. (The compressor rating was to include a 10% capacity safety factor, and the motor was sized for 110% of the compressor load.) I tried to explain to Don that we could never need a 15,000-hp electric motor. It was way too large for the capacity of the coke drums.

"Norm," Don responded, "our Mississippi refinery has a 15,000-hp compressor and we want one of the same size. After all, they have the same-size coke drums as we do."

As the Saturn refinery in Mississippi is close to my home in New Orleans, Don flew from LAX to Louis Armstrong International Airport. We drove to Mississippi. Figure 1-2 summarizes what we saw. The compressor suction valve was almost closed. The pressures were:

- *Compressor suction:* 2 psig (17 psia)
- *Compressor discharge:* 265 psig (280 psia)
- *Pressure upstream of the suction throttle valve:* 20 psig (35 psia)

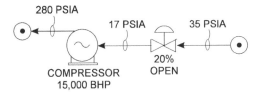

**Figure 1-2**   *Energy loss across a compressor suction throttle valve.*

Compression work per mole is proportional to the compressor suction pressure, as follows:

$$\text{compression work} \sim \left(\frac{p_2}{p_1}\right)^{0.20} - 1$$

where $p_1$ and $p_2$ are, respectively, the compressor suction and discharge pressure in psia. The exponent 0.20 is calculated from $k$ (the ratio of the specific heats) as follows:

$$k \text{ for coker wet gas} = 1.25$$
$$\frac{k-1}{k} = \frac{1.25 - 1.00}{1.25} = 0.20$$

Assume that we have two cases:

- *Case I:* suction pressure = 17 psia, suction valve 20% open
- *Case II:* suction pressure = 35 psia, suction valve 100% open

Case I requires roughly 40% more horsepower than case II. This means that approximately 30% of the motor driver horsepower is lost across the compressor suction throttle valve when it is 20% open.

## WASTING ELECTRIC ENERGY

I explained to Don that 2500 kW were being wasted at the Mississippi plant by the suction throttle valve. To generate 1 kW of electric power may require 9000 Btu/hr of fuel. In older power stations, it is likely to be 10,500 Btu/hr. Therefore, the waste in energy was

$$(2500 \text{ kW}) (9000 \text{ Btu/hr}) = 22{,}500{,}000 \text{ Btu/hr}$$

There are 6,300,000 Btu in a barrel of fuel oil. Therefore, the Mississippi wet gas compressor was wasting about 80 bsd worth of fuel a day. A typical family in the United States consumes 0.20 barrel of crude oil a day. Therefore, the suction throttle valve was wasting the amount of crude oil that 400 families would use each day.

But Don said, "Interesting, but irrelevant. Saturn has already issued the purchase order for the new compressor and motor. I'd better be careful, though, not to oversize the new compressor's suction throttle valve."

"Don, Saturn could pay a 10% cancelation fee for the compressor. The electric power saved could be exported to Los Angeles. Think of all the $CO_2$ emissions we could avoid. The concentration of greenhouse gases has been increasing at 0.5%, compounded annually, since 1975. We've got to draw the line somewhere. Why not here? Why not today?"

"Norm, I told you, Mr. Overbourne wants the new compressor."

## ALTERNATIVES TO THE NEW COMPRESSOR

"Pallas Athena, the new 15,000-hp motor was not my fault. I tried my best. There was just nothing I could do."

"Do you not fear the shades of Hades?" asked Athena.

"Not really. It's probably like the Good Hope refinery, in St. Charles Parish, Louisiana, where I used to work."

"Norman, look unto thy heart. Were there no alternatives to the new, larger centrifugal compressor?" asked the daughter of Zeus.

"Creator of Wisdom, I did have a few ideas along those lines:

- The air-cooled condensers just upstream of the compressor knockout drum were dirty. I could have cleaned their exterior to increase airflow. This would have lowered the reflux drum temperature shown in Figure 1-1. The reduced reflux drum temperature would alter the vapor–liquid equilibrium so as to produce less vapor. Also, compression work is reduced, as the compressor suction temperature is reduced, in proportion to the reduction in absolute temperature (°R or °K).

- The tube side of the air coolers could have been water-washed on-stream. Not only would this have improved cooling, but more important, the tube-side pressure drop could have been reduced. At a constant fractionator pressure, this would have raised the wet gas compressor suction pressure. Each 1 psi reduction in condenser $\Delta p$ would have reduced the suction volume by over 4%. Further, the lower $\Delta p$ value and the resulting higher reflux drum pressure would alter the vapor–liquid equilibrium so as to produce less vapor.

- The compressor could be kept clean by spraying about 1 wt% of heavy coker naphtha into the suction of the machine. Enough naphtha is used, depending on its molecular weight, to keep the compressor effluent slightly below its dew-point temperature—the idea being to keep the rotator wheels wet, which retards dry-out of salt deposits on the final stage wheels.

- Most of the time, cooler ambient conditions in Los Angeles prevent the coker from being limited by the wet gas compressor. During late afternoon summer days, water can be sprayed as a fine mist into the inlet of the forced-draft air fans. This will cool the air by 10 to 15°F due to the evaporation of the water, rather like the swamp coolers used in the south western United States. I've also seen the Russians use this technique in Lithuania and it works fine. It's a nice trick to use to reduce the flow of wet gas during peak ambient temperature periods.

- Changing from trays to beds of structured packing in the fractionator would reduce the fractionator $\Delta p$. This would raise the compressor suction pressure by an equivalent amount, but without raising the coke drum pressure.

- The interstage compressor water coolers were badly fouled. They could have been cleaned with inhibited, dilute hydrochloric acid to remove the hardness deposits on the water side. This would have reduced the vapor load to the compressor's second stage by increasing the flow of cooling water and thus decreasing the interstage knockout drum temperature. Alternatively, a chelating solution could have been circulated through the cooling water circuit, to clean the entire cooling-water circulating system.

- The air-cooled compressor discharge condenser could also have been cleaned, on both the tube and air sides. This would have reduced the discharge condenser pressure drop and thus reduced the compressor discharge pressure. This is relatively unimportant, as a 10-psi reduction in discharge pressure is equivalent to a 1-psi reduction in suction pressure as it affects the wet gas compressor compression ratio.

- The existing motor had a habit of tripping off on electrical overload. This could have been because the full-limit amperage (FLA) trip point setting was accidentally set too low. Alternatively, and more likely, the insulation integrity of the copper coil windings had deteriorated, and the motor would have to be rewound. This is not uncommon in larger motors that have seen many years of service close to their FLA point.

- Improving the cleanliness of the compressor rotor by the heavy coker naphtha spray at the suction would also help the motor. As the cleaner rotor would require less horsepower to spin, the amperage load on the motor would also be reduced.

- Another option would have been to modify the. . . ."

"Enough!" commanded Zeus.

"Merciful King of the Immortals, I sent the plant manager, Mr. Overbourne, an email with my suggestions for reusing the existing wet gas compressor and motor. But I never received a reply."

"That email! Unfortunately, it was lost in cyberspace," Zeus explained.

"But Master of Thunder, that's not my fault. I think it was returned by a Mailer Demon."

"Your sin is one of omission. You should have asked for an appointment with Overbourne. You should have insisted on your ideas. The construction of the new motor and compressor consumed resources on your little planet that will take eons to replace. Just the wasted electrical power produced more carbon dioxide emissions than Adam and Eve, who lived for 900 years," Queen Hera explained.

"Queen Hera, where's Larry Overbourne now?" I asked.

Zeus's countenance darkened, "Vengeance shall be mine. Seek for him across the River Styx, in the House of Hades."

I just hung up the phone. Don was telling me that my new fractionator and compressor designs are working great. Demonstrated capacity is overdesign and all products are on-spec.

Unfortunately, the refinery is cutting the crude run because gasoline demand is slipping due to the economic turndown. Also, the coke drums themselves are cracking due to the shortened coke drum cycles. So the extra capacity isn't needed. But still, Saturn's management is pleased with the project. I just didn't have the heart to tell Don what the ultimate top management thought about our work.

## KEEPING COMPRESSOR ROTORS CLEAN

Of the preceding list of items to improve a compressor's efficiency and capacity, the one from which I have seen the most beneficial results is the injection of a liquid spray into the suction of the compressor. This is such a beneficial practice when compressing most process and natural gas streams that I have often wondered why it is not a standard feature in most centrifugal compressor's original installations.

When I was first asked, as a young process design engineer at the old American Oil Company in 1965, to design a liquid injection system for the suction of a multistage recycle hydrogen centrifugal compressor, I was quite confused by the assignment. All compressors that I had seen were equipped with compressor K.O. (knockout) drums, to prevent liquid from damaging the compressor internals. Both entrained droplets and slugs of liquid will damage the valve plates on a reciprocating compressor. A broken valve plate will, in practice, disable the affected reciprocating compressor cylinder and lead to a loss of compression efficiency and capacity. Slugs of liquid (but not necessarily entrained droplets of liquid) will also damage both the rotor and stator of a centrifugal compressor. Hence the need for the K.O. drum ahead of the compressor's suction.

Let's assume that there is a small amount of entrained liquid in the suction of a centrifugal compressor. Actually, according to Stokes' law, that's not an assumption but a certainty. Let's further assume that this entrained liquid contains a tiny amount of salts. Again, that's not an assumption, but a certainty, if we are compressing:

- Coker off-gas
- FCU (fluid cracking unit) wet gas
- Refinery hydrotreater recycle gas
- Naphtha reformer off-gas
- Natural gas upstream of the glycol dehydration scrubber

As the gas is compressed, one might think that the higher pressure would prevent any droplets of entrained liquid from evaporating. However, the heat of compression is always a bigger factor promoting compressed gas to dry out, and it offsets the effect of the higher pressure entirely. Thus, as the inevitable droplets of liquid in the inlet gas evaporate, salts and other solids may slowly accumulate on the compressor wheels.

My most vivid experience of this common problem occurred at the Laredo compression station in south Texas in 1986. We were compressing natural gas from 600 psig to 1100 psig using a centrifugal compressor comprised of four wheels. The gas contained entrained brine (i.e., salt water). After several months of operation, I would begin to notice a gradual loss of compressor capacity. Not only would the compressor's capacity and efficiency diminish with time, but when the compressor lost about 30% of its capacity, the compressor would start to vibrate and then trip off on the high rotor vibration automatic shutdown switch. When, subsequently, the machine was disassembled, I observed that:

- The first wheel was very clean.
- The second wheel had minor salt deposits.
- The third wheel was badly encrusted with both salt and a heavy grease.
- The final and fourth wheel was very slightly fouled with salt.

Clearly, the brine was drying out on the third wheel. The resulting deposits were restricting the gas flow through the machine: thus the loss of capacity.

## CALCULATING LIQUID INJECTION RATES

A typical application in a refinery for suppressing salt formation on a centrifugal compressor's rotor, would be hydrogen recycle for a naphtha hydrosulfurizer. To calculate the amount of the liquid injection to the suction of the compressor:

- Assume that the entrainment rate is zero from the K.O. drum.
- Select the type of liquid to be employed. I would just use the naphtha stabilizer bottoms rather than an expensive specialty aromatic chemical.
- Calculate the amount of naphtha that is required in the compressor's suction to reach the dew-point temperature at the discharge from the final wheel of the stage.
- Note that each stage of compression (i.e., not each wheel) should be treated separately.
- To calculate the amount of naphtha required, take into account both the latent heat of evaporation of the naphtha and the increase in the dew-point temperature of the compressed gas, due to the gas's increased molecular weight, from the injected naphtha.
- A typical spray wash flow is 1 wt% of the gas flow. I do not add any safety factors to the amount calculated above, as ignoring the effect of entrainment in the feed gas in effect adds a safety factor to this calculation.

## DESIGN DETAILS

Then I call my Bete nozzle rep and have him select an appropriate mist nozzle for this application, I will typically specify:

- $\Delta p = 20$ psi
- Nozzle to be extractable on-stream, through a packing gland
- 316 (L) stainless steel
- External filter, with mesh openings of one-third the maximum free passage of the mist nozzle

Running without the spray while it is being cleaned for a few hours is okay. However, forgetting to shut off the naphtha spray when the compressor trips off is definitely not okay. Thus, the FRC regulating the spray wash flow should be tripped when the compressor is shut down.

I realize that changing the fractionator tray panels and adding a mist injection system to a wet gas compressor is not entirely consistent with the objective of using what we have. But these methods are a far better alternative to erecting a giant new coker fractionator tower or installing a new oversized wet gas centrifugal compressor equipped with a huge new motor, requiring a new electrical substation. This is what ethical engineering design is all about!

## TROUBLESHOOTING METHOD

In this book I detail a large number of successful examples as to how process problems were resolved, with a minimum input of new equipment. These examples are all genuine and true to life, sometimes taken out of the actual context, but without altering the technical content of the incident. However, what is not true is that I solved all these problems myself. On occasion, the solution was later found by an engineer or operator with whom I had been working.

My failure to solve some problems is usually due to spending only a day or two on a subtle or complex issue. Or, too frequently, I have made a wrong initial assumption and could not resolve the issue until this flawed assumption was discarded.

Still, almost invariably, the man or woman who has defined the correct solution after my departure has credited my troubleshooting method as contributing to their success. This method is as follows:

- Look over the equipment in the field until you become familiar with the function, location, and nomenclature.
- Discuss the problem with the hourly operators in detail.
- Carefully measure all parameters that may interact with the problem. Defining which such parameters to measure is often the most difficult and important key to solving the problem.
- Calculate the effect of process changes on the parameters.
- Observe the effect of the process changes on the parameters. That is, change things and see what happens.
- Does the observed change equal the calculated change? If not, why the difference?

The difference between the observed and predicted parameters is most often the key observation that will reveal the true nature of the problem. So often, in retrospect, the nature of the problem is disturbingly simple. Often, I've become angry at myself for stumbling over an obvious issue. But with age I've come to understand that it's my technique, not knowledge or experience, that is my fundamental contribution to the hydrocarbon process industry.

# Chapter 2

# *Vacuum Tower Heater Expansion*

"Father Zeus, what happened to H-501-A, the new vacuum heater at Saturn's El Segundo refinery?" I asked.

A section of this new, 200 million Btu/hr vacuum heater had fallen off a truck on I-10 east of Los Angeles and rolled down a hill. The damage took months to repair. It happened during a freak ice storm and was thus designated by the All Farm Insurance company to be "an Act of God."

I had been hostile to the H-501-A heater replacement project from the start. Let me explain with reference to Figure 2-1.

I had been retained by Saturn to improve the performance of its vacuum tower. The problem was excessive vertical vapor velocities in the flash zone of the vacuum tower. The high vertical velocity was promoting entrainment of black asphaltine molecules, which contaminated the gas oil product. The entrainment velocity factor, $C$, is calculated as follows:

$$C = V \sqrt{D_V \div D_L}$$

where
   $C =$ entrainment velocity, ft/sec
   $V =$ superfacial vapor velocity, ft/sec
   $D_V =$ vapor density
   $D_L =$ liquid density

*Process Engineering for a Small Planet: How to Reuse, Re-Purpose, and Retrofit Existing Process Equipment,* By Norman P. Lieberman
Copyright © 2010 John Wiley & Sons, Inc.

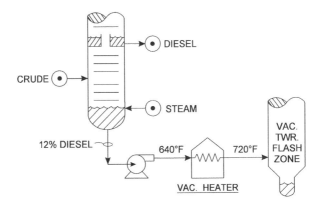

**Figure 2-1** *Diesel left in vacuum tower feed increases heater fuel consumption.*

A *C* factor above 0.45 will lead to uncontrolled entrainment. A *C* factor below 0.15 will produce very little entrainment. I design vacuum tower flash zones for a *C* factor of 0.35 ft/sec. The *C* factor calculated at the Saturn facility was quite close to the 0.45 maximum: thus the need for a complex and costly revamp.

Further, the projected feed rate for the tower was also quite high. The current vacuum tower feed analysis was very "light". Meaning, it contained a large quantity of diesel oil in addition to gas oil and asphalt. It was this lighter diesel oil that contributed most of the moles of vapor that swelled the vacuum tower's vapor flow. Also, the diesel oil had to be vaporized. I calculated the heat to vaporize the diesel oil as follows:

*Step 1.* The percent of diesel in the vacuum tower feed was 12 wt%.

*Step 2.* The total feed rate was 1 million pounds per hour.

*Step 3.* The diesel to be vaporized was therefore 120,000 lb/hr.

*Step 4.* The latent heat of vaporization of the diesel was 114 Btu/lb.

*Step 5.* The diesel had also to be heated from 640°F to 720°F, as shown in Figure 2-1. The specific heat of diesel oil is, at these temperatures, 0.70 Btu/lb-°F). This means 0.70 times 80°F (i.e., 720°F − 640°F), or an extra 56 Btu/lb of sensible heat was required.

*Step 6.* Adding together the sensible and latent heat indicates that 170 Btu/lb of diesel is required.

*Step 7.* To vaporize 120,000 lb/hr of extra diesel oil then requires about 20 million Btu/hr (120,000 × 170).

The problem was that the existing heater was limited by the combustion air rate to about 140 million Btu/hr. At the projected unit charge rates, including the excessive diesel content of the vacuum heater feed, the required vacuum heater duty was 180 million Btu/hr absorbed heat.

Larry Overbourne, the plant manager, tacked on a 10% future expansion factor, which set the new heater size at the 200 million Btu/hr of heat absorbed rated capacity. The new heater was shipped in six sections via truck. One truck skidded on an icy overpass on I-10 in West Texas, and the heater module slipped its chains and cascaded down a hillside.

"Mighty Zeus, the new heater was not my fault. I presented a detailed proposal to Mr. Overbourne, the plant manager. The proposal showed how the existing heater could have been used and the entire retrofit of the vacuum tower itself avoided."

"You did your best," said mighty Zeus. "It was not your fault. It was an Act of God."

"But why Father Zeus, Son of Cronus, why did thou cause the tray deck manways to be left out of the crude fractionation tower after the 1988 tower internal inspection?"

"It is not for man to question the ways of the gods," Zeus thundered in anger.

## MISSING TRAY DECK MANWAYS

Really! It's true! We are being tested in ways that we do not understand. For example, during the 1988 turnaround, my friend Don Davis, inspected the crude column. He insisted that all the defects on the trays were rectified. Having inspected the entire tower one last time, he approved it for final closure.

All the tray deck manways were replaced except those shown in Figure 2-2. Why the tray deck manways 21 through 24 were left off is not known. Perhaps the contractor was in a rush or had lost the required hardware. The misassembly was not discovered until four years later, in 1992, during a shutdown to install the new vacuum heater, H-501-A.

With big holes in the decks of trays 21 to 24, the internal reflux dumped straight down the tower between the diesel draw-off chimney tray and the flash zone. Contact between liquids and vapors were nil. Heavy crude oil components flashed up into the diesel product, which turned from pale yellow to murky brown. To suppress the entrainment of heavier components into diesel, the crude tower flash zone temperature, shown in Figure 2-2, was reduced from 690°F to 660°F. This increased the required vacuum heater absorbed duty for two reasons:

- The entire vacuum heater feed of 1 million lb/hr had to be heated an extra 30°F (i.e., 690°F − 660°F). At a specific heat of 0.67 Btu/lb-°F) this is

$$(30°F) (0.67) (1,000,000) = 20 \text{ million Btu/hr}$$

- The extra diesel oil content of the feed had to be vaporized at the vacuum tower flash zone temperature of 720°F. I've already calculated this in the prior section as an additional 20 million Btu/hr.

Thus, the vacuum heater required duty increased by 40 million Btu/hr because of degraded fractionation efficiency in crude tower fractionation trays 21 through 24. But the vacuum heater was limited by combustion airflow at 140 million Btu/hr, and

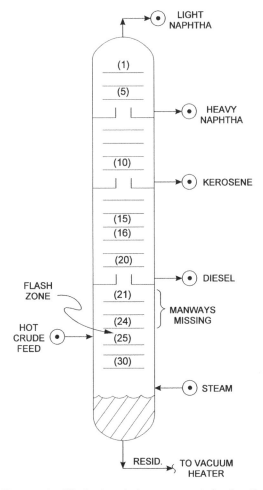

***Figure 2-2***   *Missing tray deck manways ruin fractionation.*

the extra 40 million Btu/hr heat load increased the required vacuum heater duty to 180 million Btu/hr: hence, the new heater was required.

"Master of the Thunderbolt, the overall energy efficiency of the unit was not affected," I explained. "It was a simple shift of heat absorbed duty from the crude fractionation tower preheater to the vacuum tower heater."

"No, Norman, a mortal's knowledge is imperfect, even though Eve ate of the fruit of that forbidden tree. The extra latent heat of vaporization of the diesel oil that slipped into the vacuum tower was lost to cooling water in the vacuum tower top pump-around circuit," Zeus explained. "If the diesel had been condensed at the higher pressure in the crude fractionation tower diesel pump-around, the latent heat of condensation could have been recovered by preheating crude."

"Great Immortal Zeus," said I, "grant unto me understanding."

"Poor mortal, it's simple. Diesel at 30 mmHg condenses at 250°F, too cold to exchange with crude. Diesel at 30 psig condenses at 550°F, which is an excellent temperature to preheat crude."

It's true. It's always best to recover heat from a process at the maximum possible pressure. Heat degraded to a lower pressure experiences an increase in entropy. Thus, the energy becomes less available. It's all explained in the second law of thermodynamics. For example, the value of 1 Btu at ambient temperatures is zero. The value of 1 Btu at 3000°F is its full fuel value.

## THE DIVINE PLAN REVEALED

A flash of insight! Zeus in his infinite wisdom had revealed his plan to me. It's true what the sages tell us: "Man proposes, but God disposes:"

- Don had been blinded to the omission of the tray deck manway replacement on trays 21 through 24.
- This had led to undercutting of the diesel product into the vacuum tower, which caused the vacuum heater to overfire.
- I had measured the pressure drop across trays 21 through 24 in the field. It was only 0.1 psi. I had then calculated that this same $\Delta P$ should be six times larger, 0.6 psi.
- I had thus prophetized that the four trays above the flash zone were mechanically deficient. I sent an email to Larry Overbourne with my calculations. Certainly, it would be better for our small planet to fix the trays rather than to construct a new 200 million Btu/hr heater!
- Plant manger Overbourne, forgetting that he resides on a tiny planet with limited resources, ignored the messages.
- But the son of Cronus gave Mr. Overbourne a second warning—a direct message. Zeus dropped the heater module down a hill off Interstate 10 in Odessa, and broke up the convective tube bank.

Even the Pharaoh in Egypt got the point after the tenth plague—but not Mr. Larry Overbourne. The fact that a fired heater was his downfall should have been a hint. It's also pretty hot in the shades of Hades. Not only had Overbourne caused energy to be wasted, but he also wasted natural resources. The mining of nickel and chrome for the new heater's tubes had exhausted several mines in Africa. A new natural gas well had to be sunk in Laredo, Texas for the fuel needed for the heater's fabrication shop. The concrete alone for the new H-501-A foundation destroyed an acre of land.

But I was joyous. The Immortals had revealed to me a small part of their plan. In a flash of enlightenment, my mother's words came back to me: "Norman, we're poor people. We can't afford to buy anything new. Use up what you have. Money doesn't grow on trees."

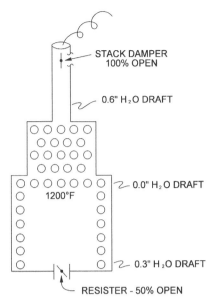

STACK DAMPER
100% OPEN

0.6" $H_2O$ DRAFT

0.0" $H_2O$ DRAFT

1200°F

0.3" $H_2O$ DRAFT

RESISTER - 50% OPEN

***Figure 2-3*** *Natural-draft-fired-heater limited by draft due to damage to convective section.*

## HEATER DRAFT LIMITATION

I mentioned that the vacuum heater was limited by combustion air. Not quite true. The heater was really limited by draft, as shown in Figure 2-3. The pressure drop across the convective section was 0.6 in. $H_2O$. If the air registers, which were halfway closed, had been opened to emit more combustion air into the firebox, the pressure below the bottom row of convective tubes would have become positive. The hot, 1200°F flue gases would have blown out of the heater box. This is very dangerous, wastes energy, and is environmentally bad news.

I measured the draft loss with a portable draft gauge. I then calculated the pressure drop on the flue-gas side of the convective section. The pressure drop calculated was 0.15 in. $H_2O$. The pressure drop observed was 0.6 in. $H_2O$, as shown in Figure 2-3.

The problem was afterburn or secondary ignition. In 1985, Mr. Overbourne had tried to save energy by overly reducing excess air in the combustion zone. This had caused excessive combustibles to enter the convective tube banks. Tramp air leaks reignited the flue gas. The resulting afterburn had caused the convective section carbon steel tubes to sag against each other. This restricted the flow of flue gas and created an excessive pressure drop through the convective tube banks.

I had sent a letter to Mr. Overbourne suggesting that he should re-tube the convective tube bank of the old H-501. Then, his vacuum heater limitation would be eliminated. But Don told me Mr. Overbourne was not interested. Why re-tube the old convective tubes when he had management approval for an entire new H-501-A heater? Don also said I had insulted Mr. Overbourne. I had implied in my letter that the damage to the convective tubes was caused by Mr. Overbourne's ill-advised 1% excess oxygen target in the heater stack.

Of course, this all happened before I realized that I was a messenger from Mount Olympus. I bet Larry Overbourne would listen to me now. You know, the shades in Hades are always looking for good technical advice on combustion problems. Hephaestus, creator of fire, has told me that Mr. Overbourne has to atone for every Btu he wasted on Earth. A big job, but he's got lots of time.

## ADDITIONAL METHODS TO INCREASE VACUUM TOWER LIFT

In Chapter 3, I offer additional ideas to increase the capacity and efficiency of process heaters by increasing air–fuel mixing efficiency. However, it isn't just an increase in vacuum heater capacity that is needed but an increase in feed rate to the crude unit shown in Figure 2-2, without an overall increase in the vacuum tower residue bottoms product. For example, there is a trade-off between the vacuum tower heater outlet temperature and the vacuum tower flash zone pressure. Figure 2-4 summarizes this relationship. Assuming that we keep constant the percent of vacuum tower feed vaporized, then the vertical axis value, "Relative TBP Cut Point, °F," is constant.

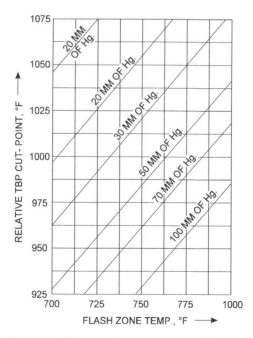

***Figure 2-4*** *Effect of heater outlet temperature vs. flash zone vacuum.*

Changes in the heater outlet temperature with the flash zone pressure can then be calculated from the family of curves and the horizontal axis. Some of the methods that can be used to lower the tower flash zone pressure are:

- Increase the bottoms stripping steam rate to the upstream crude tower (see Chapter 9). A poorly steam-stripped vacuum tower feed will overload the vacuum tower overhead ejector system.

- Minimize the operating pressure of the crude tower. This will also reduce the light ends in the vacuum tower feed. These light hydrocarbons will also contribute to overloading the vacuum jet system.

- Optimize the vacuum tower top pump-around circulation rate. If the vacuum tower has an overhead precondenser, you will find that increasing the top temperature often lowers the tower top pressure. The light ends get "sponged up" in the heavy naphtha distilled overhead.

- Reduce the motive steam pressure to the jets. Last month, I throttled back on the steam pressure to the second-stage jets on a vacuum tower in southern Louisiana. The steam pressure was reduced from 160 psig to 115 psig. This improved the vacuum by only 1 mmHg but saved 25% of the motive steam to these jets. A further reduction in the steam pressure, however, caused a loss in the vacuum.

- Back-flush the ejector condensers. Reducing the operating temperature of the condensers will always improve the vacuum to some extent. An acid cleaning of the condenser tube side will also help, as will blowing (air rumbling) the tubes out with plant air once a day.

- Check the motive steam moisture content. Wet steam will degrade the ejector performance and may actually cause a jet periodically to freeze up briefly.

- Check for air leaks. Does the vacuum tower off-gas have more than 20% nitrogen? If so, there is a major air leak in the vacuum tower or the vacuum system. An air leak into a vacuum system can be seen. When air expands into an area of lower pressure, the air is chilled. The resulting area of low temperature will cause moisture to condense around the leaking pipe or flange. Note that if the off-gas has very little $O_2$ but quite a bit of $CO_2$, the air leak is likely to be found in a hot area of the tower, not in the overhead system piping. By a "hot area," I mean the furnace transfer line in the region of high velocities.

- Listen carefully to the steam jets. Perhaps they are making a surging or "hunting" sound. This will lead to a big loss in vacuum. Often, this surging operation is caused by:
  - Overloading the jets with cracked gas
  - Erosion of the steam nozzles
  - High jet discharge pressure
  - Wet steam
  - Low motive pressure steam

## VELOCITY STEAM IN HEATER PASSES

If your vacuum tower overhead is equipped with a precondenser, high rates of coil or velocity steam in the heater passes may significantly unload the vacuum heater but without sacrificing gas oil recovery. I once lost a $50,000 contract at the Poseidon refinery in San Francisco by recommending to my former friend, Saul Pauder, that he use velocity steam in the vacuum heater passes on his unit. I say "former friend" because the coil steam was so effective in increasing gas oil recovery that Poseidon canceled my contract to revamp their vacuum tower. The coil steam has several important benefits:

- The peak temperature is suppressed in the heater passes. This happens because the steam reduces the hydrocarbon partial pressure and promotes earlier vapor-ization in the coils. The reduction in peak temperature reduces the cracked gas flow to the vacuum jets and thus reduces the flash zone pressure. Also, the lower peak temperature reduces coke formation in the tubes.
- The lower rate of coke formation in the tubes allows higher radiant heat density in the firebox and thus more heater capacity.
- The steam reduces the hydrocarbon partial pressure in the vacuum tower flash zone. This allows a lower heater outlet temperature for the same gas oil recovery rate.

If the vacuum tower has a jet rather than a precondenser on top, the use of large amounts of velocity steam will probably be counterproductive. The excessive steam flow will overload the first-stage jets and thus raise the tower flash zone pressure. Even with a precondenser, the use of velocity steam in the heater passes can be extremely counterproductive, as I learned the day I lost $50,000 during a 10-minute phone call to Mr. Pauder.

For design, a velocity steam rate of 0.5 wt% is typical. From an operating per-spective, I would use enough total steam to double the heater pass pressure drop that is observed before velocity steam is introduced.

Coil steam, of course, raises the entrainment velocity (the $C$ factor) in the flash zone. This can cause the recovered vacuum gas oil to go dark due to entrained asphaltines. Also, if the unit is limited by the vacuum heater charge pump capacity, incremental coil steam in the vacuum heater cannot be used without a reduction in the unit feed rate.

# Chapter 3

# *Natural-Draft-Fired Heaters*

Disturbed by my dream, I sought solace in religion. I began a serious study of Western religions, but none of them dealt with my problem: man's relationship to the Earth. So I decided to establish my own religion. I decided to become a messiah. To preach against the evils inherent in the use of oxygen analyzers in fired heaters. Yes, it's true. Evil lurks in the use of oxygen analyzers. Not that the analyzer itself is either good or bad; it's potentially a useful tool, like crude oil or coal. Rather, it's how we employ the $O_2$ analyzer that determines its nature. Such an instrument, installed above the heater's convective section with the objective of optimizing the combustion air rate, is a force of evil—an instrument that could have been invented by Old Nick himself.

## CONTROLLING EXCESS COMBUSTION AIR

There is no such thing as complete combustion. To have complete combustion, we would have to achieve perfect air–fuel mixing efficiency. As this is impossible, I have always observed in a heater's flue gas:

- Some excess $O_2$
- Some carbon monoxide
- Some partially oxidized hydrocarbons

*Process Engineering for a Small Planet: How to Reuse, Re-Purpose, and Retrofit Existing Process Equipment,* By Norman P. Lieberman
Copyright © 2010 John Wiley & Sons, Inc.

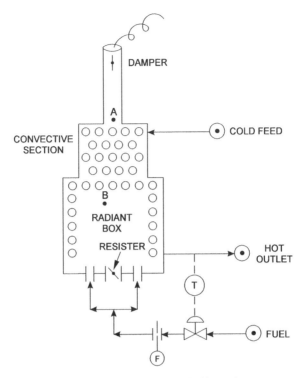

***Figure 3-1*** *Natural draft heater controlled by outlet temperature.*

It can't be helped. We can measure excess oxygen with an analyzer. But what target value of excess $O_2$ should be selected to optimize the combustion air rate? More to the point, what is meant by "optimum air rate"?

Let's assume that I'm firing the heater shown in Figure 3-1 on automatic temperature control. By "optimum air rate" I refer to that combustion air rate which minimizes the amount of fuel needed to achieve a given heater outlet temperature. I would open the air register to get more air, or close the stack dumper to reduce the airflow. Either way, I would not be guided by the irrelevant $O_2$ analyzer—I would be guided by one factor and one factor only: the flow of fuel gas. If increasing the airflow increased the fuel flow, I would use less combustion air until the fuel gas rate began to increase again.

Let's now assume that I'm firing the heater at a constant fuel gas rate. My objective is to heat the feed as much as possible with that amount of gas. How do I now optimize the combustion air rate? Once again, the $O_2$ analyzer is irrelevant. Once again I would manipulate either the stack damper or the air register to alter the airflow. But now I would watch the heater outlet temperature. The combustion air rate that maximized the heater outlet temperature is the optimum airflow.

That's how I can find the optimum combustion air rate. Not with an oxygen analyzer, but by observing my two primary operating variables: fuel consumption and heater outlet temperature. But what causes the optimum combustion air rate to be a very low $1\frac{1}{2}\%$ or a very high 6%? Certainly, the less air, the lower the heat lost up the stack with the flue gas.

It's all a matter of the air–fuel mixing efficiency. The better we mix the air and fuel, the less air I need and the more efficient the heater becomes.


## COMBUSTIBLE ANALYZERS

I could pretty much recite the same story for CO and combustible analyzers. They are a distraction from our actual objectives, which are:

- To heat the feed to the required outlet temperature while consuming the minimum amount of fuel, or
- To heat the feed as much as possible with a given amount of fuel.

What part do the $O_2$ or CO analyzers then play? They are a symbol: a symbol of our desire to save energy. They are not really needed to optimize excess air.

Okay! But can't such an analyzer be used to establish an optimum $O_2$ or CO target for the flue gas composition? Once we have found the optimum $O_2$ by trial and error, can't we just stick to that oxygen target using the analyzer as a guide?

No!—for two reasons. First, the optimum combustion air rate is a function of air–fuel mixing efficiency. Suppose that my burners are plugging because of dirty fuel gas. Then I will need more air to burn my fuel efficiently. Maybe my burner spuds (i.e., tips) are eroding out. Then what?

Second, we usually have air leaks in the convection section. The tramp air drawn into the convective box (Figure 3-1) will increase the $O_2$ content measured by the analyzer located at point A. Often, analyzers located above the convective box or the stack breeching are more a measure of convective box air leaks than of excess combustion airflow.

Could we then not locate the analyzer at point B? This is a good idea; it would certainly help. If you are really committed to the use of $O_2$ analyzers, at least locate the little devils at point B, just below the bottom row of shock tubes or the lower row of convective tubes. This will be a high-temperature installation requiring use of a relatively expensive instrument. It's a matter of degree. Oxygen analyzers located at point B are far less evil than analyzers located at point A, as shown in Figure 3-1.


## THE POWER OF POSITIVE THINKING

Why, then, are oxygen and combustible analyzers in such widespread use if they serve so little purpose? One reason is for environmental control. As air–fuel mixing efficiency gets worse at a given excess air level, such environmentally objectionable agents as:

- Aldehydes
- Ketones
- Carbon monoxide
- Light alcohols
- Unsaturated hydrocarbons
- Carcinogenic agents

are generated in increasing concentrations.

Of course, one could and should make such important measurements with an inexpensive portable unit. The real reason why on-stream analyzers are used is for political reasons. It gives management a sense of control over unit operations. It allows the technical staff to establish targets for the operators to follow, without working the problem through in the field.

Now I have to admit that there are fired heaters which do not have very many tramp air leaks, which have a constant air–fuel burner mixing efficiency, and where the fuel gas composition is constant. In such a case, I'll agree that an oxygen analyzer is quite a useful tool. But if you work in a complex refinery, with a variable fuel gas composition, well—you have to wonder. You have to wonder which salesperson sold plant management 16 oxygen analyzers that can only be installed in the stack breeching, because they have a low temperature rating. Probably a salesperson following in the steps of Dr. Faust or bargaining with some other evil entity: the evil spirit that is always trying to promote useless projects for personal financial gain.

Why do my methods represent truth and morality? Why are my methods smiled upon by Queen Hera, the Earth Mother, and Pallas Athena, Goddess of Wisdom? Because you can use my methods with what you have. You can start today. Mainly, though, you can follow my mother's wise advice. "Norman!" my mother would shout, "We're poor people! We can't afford any complicated combustible analyzers. Get by with what you have. Get out of the house. Get some fresh air. It's a nice day outside. The sun is shining. Get out of the office and adjust some air registers. The fuel gas rate will go up or down. If it goes down, crank down on the air registers a bit more. Why can't you be more like your sister? She has so many friends. She always...."

But I'm getting my metaphors confused. Perhaps I had better relate how to improve the burner air–fuel mixing efficiency, to reduce the combustion air requirements with the objective of saving fuel, thus reducing $CO_2$ accumulation in the atmosphere.

## IMPROVING THE AIR–FUEL MIXING EFFICIENCY

The purpose of the burner is not to burn the fuel. The purpose of the burner is to mix the fuel and the air. This means that combustion air needs to flow through a burner that is actually lit. Walk underneath a heater with floor-fired burners. Make a list of:

- Jammed secondary air registers on burners not in service that cannot be shut
- Site ports with missing glass covers
- Pilot light ignition ports without their covers
- Burners not fitted tightly in place
- Miscellaneous cracks and crevices

All these openings allow combustion air to flow into the fire. But as this airflow bypasses the burners, it does not mix efficiently with the fuel. The fuel gas then wanders around the firebox searching for air. Long, licking, yellow or orange smoky flames fill the firebox. The flames impinge against the roof and upper wall tubes. If you don't wish to overheat tubes due to flame impingement, you had best open the secondary air registers. This will bring the flames back closer to the burners and away from the tubes, which is good for the tubes but bad for the planet. The extra airflow generates more flue gas, which increases the heat loss up the stack, which reduces heater efficiency. This requires more fuel and generates more greenhouse gas in the form of $CO_2$.

Alternatively, you can fix all the tramp air leaks in the radiant firebox, including the sight doors hanging off their hinges on the side of the heater. On the other hand, if you order an oxygen analyzer to save energy, you will have to:

- Justify the expenditure.
- Get the funds appropriated.
- Go out for competitive bids.
- Have the purchasing agent do his job.
- Argue with the maintenance department to install it.
- Write procedures for its use.
- Train the operators, who don't like you.

By the time you are halfway done, you'll be promoted, or fired, or transferred to Indonesia. If you use my methods, you can drop this book, rush out to the plant, and start saving energy right now!

## CONVECTIVE SECTION TRAMP AIR LEAKS

As I write this, I'm flying home from a refinery in Kansas. I was working on a crude fired heater that was limited by the burner tip pressure, meaning that it could not fire more fuel. I had the operators pinch back on the stack damper to diminish the draft. This raised the pressure at point B in Figure 3-1, from minus 0.30 in. $H_2O$ to minus 0.10 in. $H_2O$. With an increase in the convective box pressure, the flow of cold tramp air entering through the holes in the convective section was reduced. The convection then ran hotter. More heat was transferred to the crude oil charge without increasing the fuel consumption. The heater outlet temperature rose from 698°F to 702°F. The refinery recovered an extra 400 bsd of diesel oil from reduced crude, which is what petroleum refineries are supposed to do.

I'm going home. I've had a glass of wine, and I feel really good about the entire situation. Like a walk in the forest, I'm at peace with nature.

And I have other, simpler methods to stop air leaks in the convective box, in addition to reducing draft. Get a box of baking powder. Let it drift in front of any potential air leaks in the convective section box, especially where the seams are bolted together. Most convective boxes leak. When you find these tramp air leaks, use silicon sealer or insulating mud or a roll of aluminum duct tape to fix the leaks.

"Better to light one small candle then curse the darkness."

To evaluate your progress, measure the oxygen content at both points A and B (Figure 3-1) using a portable $O_2$ analyzer. Each 1% $\Delta O_2$ reduction will save about 1.2% of the furnace fuel fired (assuming a 60°F ambient air temperature, a 660°F stack temperature, and 15% excess air).

Closing the stack damper is fine up to a point. That point is reached when the firebox develops a positive pressure. Then the flue gas containing $SO_2$ (sulfur dioxide) will blow out of the convective section leaks and radiant section site ports. If you're firing 6% sulfur heavy industrial fuel oil as we did at the Carib refinery, the $SO_2$ will, in all probability, kill you. This is no big deal. Don't worry! No one lives forever anyway. The important thing is that you have died in nature's service. Trust me. Just tell the angels during your exit interview that you were minimizing the convective section tramp air leaks to save energy and thus reduce greenhouse gas emissions in the form of $CO_2$.

## AIR PREHEATERS

As part of my new religion, I also preach against the evils of air preheaters. Air preheaters, if properly designed, typically may save 5 to 10% of a furnace's fuel consumption. Figure 3-2 shows the sort of an air preheater installation that I oppose, and Table 3-1 illustrates the problem of tube leaks in an air preheater.

**TABLE 3-1   Effect of Tube Leakage on Air Preheater**

|  | Design | Actual |
|---|---|---|
| Flue gas to air preheater | 600°F | 600°F |
| Flue gas from air preheater | 320°F | 240°F |
| Combustion air to preheater | 20°F | 20°F |
| Combustion air from air preheater | 320°F | 260°F |
| $O_2$ in flue gas to air preheater | 2% | 0.5% |
| $O_2$ in flue gas from air preheater | 2% | 9% |

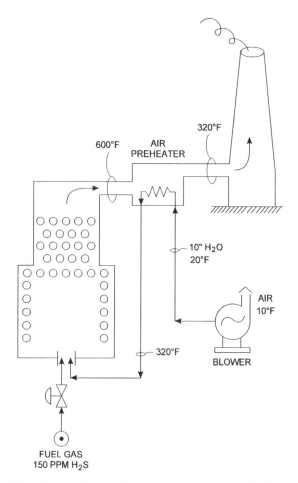

**Figure 3-2**   *Air preheater is subject to tube leakage due to $H_2SO_4$ corrosion.*

When we burn fuel gas or fuel oil, most of the sulfur is converted to $SO_2$, which is not corrosive. A smaller portion of the sulfur is converted to $SO_3$, which is also not corrosive. The $SO_3$ reacts with water formed from combustion of hydrogen in the fuel, to form sulfuric acid, which when dry is not corrosive. But depending on the sulfur content of the fuel and the flue gas temperature, sulfuric acid will precipitate out of the flue gas as a sulfuric acid mist. This acid mist is most certainly corrosive to the air preheater tubes shown in Figure 3-2.

Let me give you several potentially misleading parameters:

- If you are burning clean refinery fuel gas with less than 100 ppm $H_2S$, the sulfuric acid should not form a mist at temperatures over 280°F.
- If you are burning 5% sulfur fuel oil as we did in Aruba, sulfuric acid should not form a mist at temperatures over 450°F.

These theoretical temperatures for $H_2SO_4$ mist precipitation are meaningless because they fail to take into account the localized cold wall temperatures of the air preheater tubes, due to the fact that the air being preheated is at 20°F. The localized cooling causes localized tube leaks due to $H_2SO_4$ attack, which forms $FeSO_4$ (iron sulfates). These tube leaks then promote the following nasty problems:

- Cold air flows from the higher combustion air pressure into the lower flue gas pressure.
- The cold air causes more cold areas to form in the flue gas, which promotes more $H_2SO_4$ precipitation plus more corrosion.
- The cold air leaking into the flue gas quenches the flue gas and diminishes air preheater and energy efficiency.
- The resulting diminished stack temperature reduces the draft in the heater.
- The accumulation of $FeSO_4$ on the flue-gas side of the air preheater increases the air preheater $\Delta P$ and also increases backpressure on the heater. This further diminishes the heater's draft.
- With combustion air blowing right up the stack, the heater itself becomes combustion air limited. The loss of draft just makes the lack of combustion air worse.

I have never seen an air preheater (Figure 3-2) working for more than a few years before it develops terrible tube leaks. By "terrible" I mean that the majority of the air blower discharge is blowing right up the stack. The problem is worse in colder climates and in plants that have $H_2S$ excursions in their fuel gas.

In practice, it's the sulfate deposits in the flue gas side that cause a furnace-limiting positive pressure to develop at point B in Figure 3-1. As an engineer who professes to be promoting energy conservation, I have often sinned by opening the flue gas bypass around the air preheater to relieve the resulting heater draft limitation. I did this once for a crude unit heater in Convent, Louisiana, and "de-bottlenecked" the crude rate by 10%.

## CORRECT AIR PREHEATER DESIGN

The overriding problems with theoretical vs. real-world air preheater design is that even in southern Texas and Louisiana, ambient conditions sometimes drop to 20°F. Also, even in refinery fuel gas that is supposed to run at no more than 150 ppm $H_2S$, excursions of 10,000 ppm $H_2S$ happen. The problem is that once an air preheater tube leak starts, a positive feedback loop is created, and the leakage feeds on itself.

The air preheater at the No. 2 crude unit at the Ever Hopeful refinery prevented these problems. We used a circulating hot oil stream to recover heat from the convective flue gas. Then, in a separate box, the hot oil preheated the combustion air

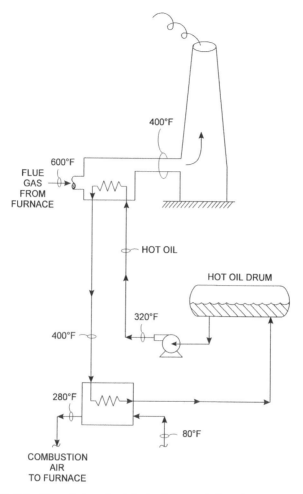

***Figure 3-3*** *Correct air preheat design to prevent air preheater tube leaks.*

(Figure 3-3). Because of the indirect heat exchange via the hot oil circuit, this air preheat method will:

- Never exchange heat as efficiently as the direct heat exchange shown in Figure 3-2.
- Probably cost twice as much to purchase and install as will the direct heat exchange design.
- Have a higher power consumption, due to the hot oil circulation pump.
- Not promote tube leaks in the air preheater.

Alternatively, I have seen refineries in France preheat cold combustion air using low pressure (1 bar) steam to 50°C. The preheated air, combined with a consistent and very low $H_2S$ content in the fuel gas, seems to greatly retard air preheater tube leaks.

## REDUCING HEATER CAPACITY USING AIR PREHEAT

If you preheat combustion air from 40°F to 400°F, you will save about 10% of the required fuel. However, the temperature of the flames will increase and thus the firebox will run hotter. Since most heaters are limited by radiant heat density, bridge wall temperature, or a high firebox TI, the operators will reduce the fuel consumption in the heater as a consequence of the air preheater. Then the flow of convective section flue gases goes down and the heat pickup in the convective tubes (Figure 3-1) between points A and B is reduced. This loss in convective section heat transfer diminishes the overall heater capacity by about 3%. That's a price we have to pay for energy efficiency. Of course, you can offset this lost capacity by increasing the combustion airflow by 10% to generate more flue gas flow and thus more convective section heat duty.

Delayed coker heaters seem especially sensitive to higher adiabatic flame temperatures. Thus, such services may not be good candidates for air preheaters, due to coke formation inside the furnace radiant section tubes.

My negative attitude toward oxygen and combustible analyzers, as well as air preheaters, may seem regressive to many readers. But my experience teaches that some sorts of technology have to be used carefully; otherwise, they can do more harm than good. Maybe in 100 years, our entire fossil fuel, energy exploitation economy may in retrospect appear as shortsighted and ultimately destructive to human civilization.

# Chapter 4

# Crude Pre-Flash Towers

I was attending the API annual meeting in New Orleans. I hadn't registered, only dropped by for the free snacks. In the lobby of the Orleans Hotel, I saw Glen, a process engineer with the Orion refinery, talking to an attractive girl.

Glen introduced me to Diane, also a process engineer. Sparkling blue eyes. A red dress. Perfect skin, golden hair. She was exactly the type of girl who had ignored me since high school.

"Are you really Norm Lieberman, the famous engineer? I never thought I would meet you in person! But I was just leaving. Can I buy you a coffee?"

"Okay," I said. "I'm just leaving, too."

Over coffee, Diane explained: "Norm, this is a stroke of luck. I've my own engineering company. We do process design revamps. Do you know much about crude unit pre-flash tower tray flooding?"

"Sure, Diane. I've written papers on that topic."

Diane smiled. Her teeth were a dazzling white. "I have to admit that I've read everything you've written. I'm very interested in you, Norm."

"Really?" I asked.

"Yes, really, Norm. Just today I signed a contract with the Jupiter refinery to revamp their crude pre-flash tower. It's flooding and making black naphtha."

"That's bad, Diane."

"No, Norm! It's good for us. Jupiter wants a process design for a replacement pre-flash tower. The process design engineering fee is $280,000. I have the signed

*Process Engineering for a Small Planet: How to Reuse, Re-Purpose, and Retrofit Existing Process Equipment,* By Norman P. Lieberman
Copyright © 2010 John Wiley & Sons, Inc.

contract." She reached across the table and touched my hand. "It will be great working together."

"Diane, I don't know. I dislike splitting engineering fees. What's my percentage?"

"Norm, I only got the contract because I promised Jupiter Oil that you would be doing the design. We'll split the fee: $10,000 for me and $270,000 for you."

"But, Diane, how did you know I'd be here. . .?"

"Oh! I saw your name on the registration list. You know, Norm, my interest in you goes beyond business," Diane said, as her knee brushed against mine.

"But I didn't register."

"Look," Diane said firmly, "Here's the P.O. from Jupiter Oil. I have half your payment ready. Let's both sign the agreement." She really did have a purchase order from Jupiter Oil for $280,000. The contract was ready for our signatures. Diane had a cashier's check for $135,000.

She extracted a gold pin from her purse. "Let's sign in blood," she laughed, as she pricked her finger and traced her name on the contract. "Now you!"

I also pricked my finger with the golden pin and signed. Diane smiled and handed me the check. "Got to run, Norm. My contact info is on my card. Call me when the tower design is finished."

## PRE-FLASH TOWER FLOODING

I had my doubts about this project right from the start. The problem was the black naphtha. Tray flooding is a consequence of two possible problems:

1. Lack of tray hydraulic capacity. This could be due to downcomers being too small. Or excessive vapor velocities. Or excessive weir loading. Or a too small tray deck vertical spacing. Possibly the tray hole area was less then optimum (i.e., 15% maximum).
2. High tower bottoms level.

As the symptom of the problem was black naphtha, I was quite sure that it was a high liquid level that was the culprit. To explain, let's refer to Figure 4-1. Note that the hot crude enters below tray 5 at 350°F. The hot vapors flow upward from tray 5 and are cooled off by the downward-flowing reflux. The upward-flowing vapor is cooled largely by the evaporation of the reflux. It's rather like cooling a sick child with an alcohol rub. The cooling effect is due to the evaporation of the alcohol.

I might say that I am converting the sensible heat content of the hot child's skin to latent heat of evaporation of the alcohol. Or, I might say that I am converting the sensible heat content of the up-flowing vapors to the latent heat of evaporation of the down-flowing reflux. Then, the pounds of reflux so vaporized adds to the pounds of vapor flashed from the hot crude. Therefore, the mass flow of vapor is greater on tray 1 than on tray 5.

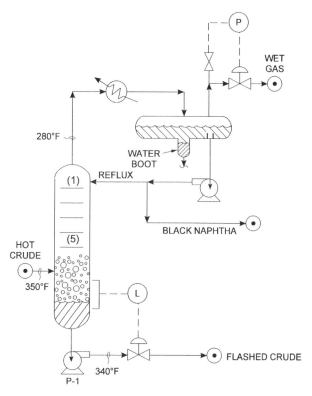

**Figure 4-1**   *Crude pre-flash tower suffering from foam-induced flooding.*

Also, the molecular weight of the vapor flow is also lower on tray 1 than on tray 5. As vapor volume is inversely proportional to molecular weight, the volumetric flow of the vapor is also greater on tray 1 than on tray 5.

It's true that the 70°F reduction in vapor temperature across the trays somewhat offsets the effect of evaporating the reflux. But this is a relatively small effect. The net effect of the reflux in the crude pre-flash tower is to increase the vapor flow rate from tray 5 to tray 1 by roughly 20%.

Also, the liquid rate leaving tray 1 is about three times higher than that of tray 5 because most of the reflux evaporates as it cools the up-flowing vapors. Thus, both the vapor load and the liquid load are larger on tray 1 than on tray 5. Therefore, the top few trays of a pre-flash tower will flood due to tray hydraulic limitations before the bottom few trays. The result of this flooding will be high-endpoint naphtha rather than black naphtha.

By "high-endpoint naphtha," I mean that the overhead product is contaminated with heavier, higher-boiling-point naphtha molecules but not with entrained, black crude oil components. Thus, tray flooding due to excessive vapor and liquid rates should never result in black naphtha.

However, if tray flooding occurs due to high liquid levels, then the black crude oil components will be entrained up the pre-flash tower trays. If the tower's bottom level (see Figure 4-1) rises above the feed inlet nozzle, entrainment of the flashed, black crude oil is certain to result.

You will recall that Diane told me that the Jupiter pre-flash tower was producing black naphtha, not high-endpoint naphtha. This statement by itself proved that the problem was not lack of tower tray capacity but too high a tower's bottom liquid level. Diane probably had not actually read my article, "Foam Induced Flooding," published in *The Oil and Gas Journal* [1], in which I explained how foam in the bottom of any tower causes the real level in the bottom of the tower to be higher than the level indicated on the control panel. Thus, the level control needs to be calibrated to account for the reduced density of the foam at the tower bottoms (see Chapter 16).

Flashed crude has a distinct tendency to foam at about 350°F. Thus, an indicated level on the control panel of only 60% may actually correspond to a froth level in the tower bottoms above the feed inlet nozzle. The result would then be entrainment of the flashed crude oil and black naphtha.

With these thoughts in mind, I drove out the next day to Jupiter Oil, located in Chalmette, Louisiana, which is adjacent to my home in New Orleans. I asked the operators to reduce the pre-flash tower bottoms level set point from 60% to 40%. Within an hour, the color of the overhead naphtha product turned from a murky brown to a clear, pale yellow.

I monitored the tower's improved operation for several hours. But shortly after lunch, the naphtha turned black again, even though the panel board indicated that the level was still holding on automatic control at a 40% level. Now what?

I drew a sample of the murky naphtha product in a quart bottle. I noticed a small amount of water in my bottle that I had not observed in prior samples. The reflux flow was also contaminated with water. Apparently, we had started to reflux water back to the column (see Figure 4-1).

Certainly, permitting free liquid water to mix on trays with liquid hydrocarbons is going to create a foamy emulsion of oil and water. As the foamy emulsion works its way down the tower, it will cause all the trays to flood, including tray 5. Flooding of tray 5 will promote entrainment of black crude oil components into tray 1. This will turn the overhead naphtha product black.

I now checked the water boot level shown in Figure 4-1. The water-level glass was full. I blew out the level control taps on the boot, which had plugged with iron sulfide scale. The dirty water level in the boot dropped by 2 ft. Half an hour later, I found that a new sample of reflux was free of water. After another half hour had elapsed, the overhead naphtha product was clear and bright again.

I explained to Robert Glass, the plant manager at Jupiter Oil, that the pre-flash tower was adequately sized for the crude rate. The problems were simply:

1. A high tower bottoms level due to the tendency of the flashed crude to foam. This problem was dealt with simply by reducing the tower bottoms level, consistent with not cavitating the flashed crude booster pump (P-1) shown in Figure 4-1.

2. Plugged level taps on the reflux drum water draw-off boot. The level taps should be blown out once a day to preclude refluxing water onto the trays, which caused emulsion formation on the tray decks.

Mr. Glass was upset. As a matter of personal pride, he had wanted a new pre-flash tower. The old pre-flash tower was a dirty, rusting structure. He had budgeted $4 million for the new tower and $280,000 just for the process design.

I looked at my watch. It was 3:30 P.M. I had been in the plant since 8:30 A.M. I wrote out my invoice for seven hours consulting, totaling $1980. Mr. Glass looked at my invoice with disgust. "Give it to Diane," he said, obviously annoyed and disinterested. "She has the contract for this project."

## THE WAGES OF SIN

So I drove out to the address on Diane's business card on St. Charles Avenue.

"So, Norm, finished so soon? Do you have the vessel sketch for the new pre-flash tower?" she asked.

"Not exactly. I'd better return the cashier's check. But you owe me $1980. Actually, Jupiter Oil may cancel the project. The tower's working fine now. It was just a couple of level control problems, which I fixed."

"And where, Mr. Lieberman, in our contract does it say anything about 'fixing problems'?" Diane asked angrily.

"Really, Diane, Mr. Glass said he would pay you 10% of your contract price as a cancelation fee. I've returned your $135,000 check. What's the big deal?"

Diane had aged. Same blonde hair. Same pearl necklace. She looked the same, yet somehow different. Probably just the lighting in her office.

"It's not the money," she hissed. "It's the principle."

"What principle," I asked defensively. "I just think avoiding a useless project is a good thing. That's what engineering ethics are all about."

"Engineering ethics!" she positively screamed. "Who are you to talk about 'ethics'? How about the contract you signed in blood? Where's the design for the pre-flash tower vessel that you contracted to produce? For which you accepted payment in advance. You signed the contract of your own free will. You entered into an agreement with me."

"Diane, can I tell you about this dream I had last month? I dreamt that I had. . . ."
"No," she interrupted, "I know all about that dream. That's the reason I'm here."

The beautiful girl with the red dress had been transformed. Her luminous skin had turned wrinkled and pale. Her slanted eyes glowed with fury. Now I understood. I had sold my soul to a she devil for a $135,000 cashier's check.

"Yes, Norman, we know all about that dream," Diane said in a more composed voice. "Don't think for one moment that you 'Save the Earth' types will succeed. We'll get that new pre-flash tower built. Mr. Glass at Jupiter Oil has signed up with us. Don't think even for one minute that the idea of field troubleshooting will stop our vast array of impressive new projects. We're going to build more new polypropylene plants and cat crackers. We will stuff the planet with cars, planes, and plastic bottles. And we'll do it without you. We're going to exploit oil shale, tar sands, and LNG in Qatar, and convert coal to oil. We will start with heavy oil. . . ."

But this I barely heard, as I escaped to St. Charles Avenue. Now, just the suggestion of designing new distillation towers makes me shudder with fear. The wages of sin are high, but Queen Hera is watching. The moral of this story is that the forces of evil want new process equipment, but Queen Hera and Pallas Athena want us to use what we have.

## ENERGY SAVINGS WITH PRE-FLASH TOWERS

The pre-flash naphtha is recovered at a much lower temperature level, than the naphtha in the crude tower. The actual energy savings would be very difficult to calculate rigorously, and would be unique for each application, depending on the pump-around heat exchange configuration. Regardless, my rule of thumb for predicting the energy benefits at the pre-flash tower is

$$\Delta H = (T_2 - T_1) (0.6) (\text{lb}) \tag{4-1}$$

where

$\Delta H$ = reduction in the crude heater absorbed duty, Btu/hr
$T_2$ = temperature in crude tower flash zone, °F
$T_1$ = temperature in pre-flash tower flash zone, °F
lb = pounds per hour of pre-flash naphtha (excluding reflux)

Note that a high top pre-flash tower reflux rate will largely eliminate the energy incentive calculated above (see Chapter 9).

## CAPACITY BENEFITS OF PRE-FLASH TOWERS

Guy Walker, my New York friend and office mate, and I designed the retrofit of No. 12 P.S. in Whiting, Indiana in 1965. This was before the computer age, and the calculations required several months. Now the calculations require several minutes. Still, Guy Walker achieved an expansion of the crude charge rate of 30,000 bsd (from 180,000 bsd to 210,000 bsd), mainly by reducing the vapor velocity in the flash zone of the crude column.

Guy Walker cleverly installed five trays (see Figure 4-1) in the top of the existing pre-flash drum. Reflux was used to control the naphtha endpoint so as to minimize heavier naphtha components (i.e., benzene precursors) in the overhead naphtha product.

The average molecular weight of the crude wet gas and the lighter naphtha distilled overhead was about 80. The pre-flash tower, which is located downstream of the crude desalter, also flashed overhead the residual water from the desalter, which typically runs at 0.3 to 0.5 wt% of the crude charge. The average molecular weight of the heavy naphtha, jet fuel, diesel, and light gas oil vapors in the crude tower flash zone is about 200. So that even though only about 10 vol% of the crude charge is recovered from the pre-flash tower overhead, the reduction in the vapor load, which is largely a function of the number of moles of vapor in the crude tower, can exceed 25%.

The reduction in the flash zone molar vapor flow also permits a proportionate reduction in the overflash or wash oil rate, on the trays beneath the atmospheric gas oil draw-off product. This also increases the gas oil recovery in the crude tower.

On the negative side, the removal of light naphtha and wet gas from the crude tower flash zone will make it more difficult to flash diesel oil and light gas oil up the tower. The moles of lighter components help vaporize the diesel and gas oil moles. This can be offset by increasing the flash zone temperature. This factor does not actually cause much of an increase in the heater firing, as the extra heat is recovered as:

- Higher feed temperature to the downstream vacuum heater (i.e., crude tower bottoms).
- More diesel oil and gas oil pump-around heat recovery from the hotter product draw temperatures.

Predicting the exact benefits of a pre-flash tower is really a very complex problem, especially when the nature of the crude is variable. The lighter the crude, the more beneficial will be the pre-flash tower. For me, I will simply use equation (4-1) and take the result as a percentage of the heater duty. Then I assume that percentage will more or less represent the incremental capacity obtained by retrofitting a crude unit with a pre-flash tower. My experience indicates that a pre-flash tower is often the most cost-effective and energy-efficient method available to run incremental crude. Also, it's an excellent method to use to enhance the energy efficiency of any crude unit.

The main drawback that limits the benefits of the pre-flash tower shown in Figure 4-1 is the top reflux. However, I have dealt with that difficulty as well, as I show in Figure 9-1, which illustrates a concept I developed in 1969 for using the pre-flash naphtha product as an intermediate reflux on the upper trays of the crude tower.

In summary, I am a big fan of crude pre-flash towers and have observed their beneficial energy and capacity effects at many refinery locations. The energy benefit in particular may be enhanced by the "intermediate reflux" concept. The optimum location for the intermediate reflux return to the crude tower is a complex problem requiring the use of a computer model of the unit.

## PRE-FLASH TOWER EXTERNAL REFLUX

Guy Walker and I shared an office for four years. He was a devout deer hunter. He spent many an hour explaining the benefits of slaying deer. I think he was trying to convert me to hunting.

I, on the other hand, expended much effort explaining to Guy Walker the benefits of pre-flash tower external reflux. Neither of us succeeded in achieving our objective. More significantly, the pre-flash tower that Guy and I designed for the Whiting refinery was constructed with a complete and conventional overhead system consisting of:

- Overhead water-cooled condensers
- A reflux drum, with a water draw-off boot
- A reflux pump
- A water boot pump

"Guy," I would explain, "I've got my principles, too."

"Norm, what principles?" Guy countered. "When's the last time you went hunting?"

"About 12 years ago, but that's irrelevant. My principles are based on minimizing construction of new process equipment. I don't think we have to build a new overhead system for the pre-flash tower. We can use the existing overhead system."

"What's the overhead system got to do with hunting? Where do you get these crazy ideas from anyway, Norm?"

"Guy, my mother said we live on a small planet with limited resources. My mom was kind of religious."

"What's your mother got to do with our pre-flash tower design?" Guy asked, obviously annoyed.

"Look, Guy. My mom said we're poor people and we must use what we have. It's against my principles to build a new pre-flash tower overhead system My mother came to me in a dream and inspired me to design the overhead system shown in Figure 4-2."

"Your mother inspired you? Dr. Horner will never approve your mom's idea of cross-connecting the two towers."

"Guy, your attitude toward my mother is rather hostile. I'm sure I would treat your mother's ideas with greater respect."

"Okay, Norm, let's submit your shared overhead system design to Dr. Horner as an alternative design case. But it seems to me that we're going to be overloading the existing crude tower condenser (C-1 in Figure 4-2), with the extra vapor generated in the pre-flash tower."

"No, Guy. Not true! The vapor load to the C-1 condenser will be the same with or without the pre-flash tower. It's just that the vapor from the pre-flash tower is bypassing those trays in the crude tower below tray 3."

But Dr. Horner, the V.P. of engineering, rejected my design on the basis that he was the vice president and I was not. But in 1981, I retrofitted the No. 1 crude unit at the Forever Hopeful refinery in Louisiana by adding trays to the existing pre-flash drum and cross-connecting the overhead system of the crude and pre-flash drum as I had proposed to Dr. Horner many years before. And it all worked out just fine.

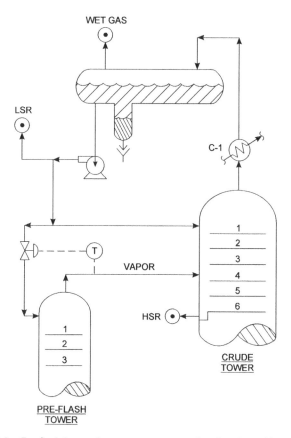

**Figure 4-2**    *Pre-flash tower shares a common overhead system with crude tower.*

Caution! The downside of the arrangement shown in Figure 4-2 is that should black crude oil components become entrained in the pre-flash naphtha vapor, all the distillate products from the crude tower (HSR, jet, diesel, AGO) will turn brown.

## REFERENCE

1. Lieberman, N. P. "Foaming Is Leading Cause of Tower Flooding," *Oil and Gas Journal*, Apr. 14, 1989, p. 45.

# Chapter 5

# Amine Regeneration and Sulfur Recovery

With autumn closing in, after my brief affair with Diane ended, I felt a longing. Not for Diane, but for the $135,000 cashier's check. Autumn in New Orleans marks the end of the oppressive heat. Thus, I started back to work with enthusiasm for this new season in my life.

Another factor accounted for my confidence. My new religion was going well; I had made my first convert: the technical service manager, Josh, at a refinery in Texas. Josh came in third in the long jump in the 1972 Olympics, representing Ghana. The refinery was owned by an Israeli company, Shalom refining. Josh reasoned that a Jewish boy from Brooklyn would be appreciated by the new refinery owners. Also, he had sulfur emission environmental problems and only a small budget available to correct them.

Josh said, "Norman! I heard about your new religion! That we, as refinery engineers, are supposed to expand capacity and improve efficiency by the power of positive thinking. Without any capital investment. Does this also apply to sulfur recovery and amine systems? I've got an awful excess of $H_2S$ to convert to sulfur! Can your new religion help me recover more sulfur?"

"Josh, we can certainly minimize capital investment. But there is my process design engineering fee of $300 per hour," I answered.

"But can I count that as a tax-deductible religious contribution?" Josh inquired.

"Yes, Josh. I am fundamentally opposed on moral grounds to building new amine regenerators or new sulfur recovery units. If you can make a $135,000 donation to my temple fund in New Orleans, I will, with divine guidance, provide you with process

*Process Engineering for a Small Planet: How to Reuse, Re-Purpose, and Retrofit Existing Process Equipment.* By Norman P. Lieberman

engineering details to expand $H_2S$ recovery by 30%. In return for your contribution, I will supply you with PFDs and marked-up P&IDs for the expansion."

"Norm, how about dropping a zero? Let's call it $13,500."

"All donations gratefully received," I sighed.

## AMINE CAPACITY EXPANSION

Even the $13,500 was a big donation to my temple. I already knew exactly what to do. To start, I asked Union Carbide, which supplied the DEA (diethanolamine), to measure the heat-stable salt content (HEED) in the circulating amine. The lab analysis reported 2.6 wt% heat-stable salts in the circulating amine. The total circulating stream was 28.2 wt% DEA. So the first step was to remove the heat-stable salts. This is done with a rented ion-exchanger unit. There are several companies based in Calgary that supply this service. Amine is circulated through the skid-mounted unit for several weeks, depending on the volume of the amine system. It might cost $50,000 to $100,000 to reduce the heat-stable salts from 2.6 wt% to 0.5 wt%. But the amine circulation rate will also be reduced by

$$\frac{(2.6 - 0.5) \text{ wt\%}}{[28.2 - (2.6 - 0.5)] \text{ wt\%}} = 8.1\%$$

If the amine circulation rate is 1000 gpm, steam savings on the amine regenerator reboiler (see Figure 5-1) would be about 4000 lb/hr, which is worth about

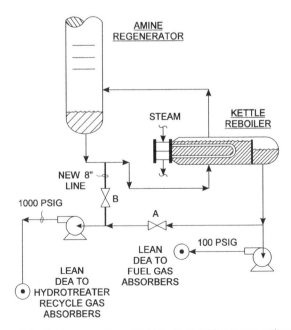

**Figure 5-1** *Saving capacity and energy in an amine regeneration plant.*

$1000 per day, depending on the value of the 50-psig steam to the plant. From an environmental perspective, less steam consumption results in less $CO_2$ emission.

What else can we do to reduce the amine circulation rate? One trick is to reduce the lean amine feed point location on the fuel gas absorbers, scrubbing $H_2S$ from the cat cracker and delayed coker off-gases. These streams will have substantial amounts of $CO_2$ as well as $H_2S$. An absorber with 20 trays may absorb 99.95% of the $H_2S$ and 90% of the $CO_2$. Using only eight of the trays will reduce absorption to perhaps 99.9% of the $H_2S$ and 60% of the $CO_2$. Depending on the ratio of $CO_2$ to $H_2S$, a significant quantity of amine circulation may be saved. The required amine circulation rate is proportional to the total ($H_2S$ plus $CO_2$) acid gas absorption, not just $H_2S$ absorption. At the Shalom refinery in Texas, the concentration of $CO_2$ was 1% and the concentration of $H_2S$ was 4% in the FCU fuel gas. About 60% of the refinery amine circulation was used to scrub $H_2S$ from this fuel gas. Relocating the lean amine nozzle 12 trays lower in the absorber would reduce the amine circulation rate by

$$\frac{1\%}{1\% + 4\%} (60\%)(90\% - 60\%) = 3.6\%$$

Another method is to increase the circulating amine strength. For DEA, an acceptable strength for circulating clean solution is 35 wt%. At the Shalom refinery, circulating strength was limited to 28 wt% based on the tendency of the contaminated amine to foam due to surfactants and particulates. The amine was greenish gray. Particulates, in the form of iron sulfide corrosion products, were supposed to be removed in the slip-stream cartridge filter. However, the cartridge filter was not assembled properly, and thus no particulates were being removed.

I knew this for sure. The $\Delta P$ value across the filter was zero. Also, the amine flowing into and out of the filter was equally dark gray. I had learned a trick while working as an operator during the 1980 strike at the Amoco refinery in Texas City. I hated to disassemble the filter housing and pull out the filthy cartridges. By accident, I once did not reinsert two new cartridges carefully. Then the filter never had any increase in $\Delta P$, due to internal bypassing. Apparently, someone at this Shalom refinery had made a similar discovery. After 10 days of proper particulate filtration and the consumption of a hundred 48 in. $\times$ 3 in. cartridges, the amine was a pale, clear, greenish color and free of particulates. I also had the carbon in the charcoal carbon filter renewed. The charcoal filter removes dissolved impurities such as heavy aromatics and spent defoaming agents that act as surfactants.

Having suppressed the foaming tendency of the circulating solution, we could increase the amine strength from 28.2% to 33%. This would reduce the required amine circulation rate by

$$\frac{33.0\% - 28.2\%}{33.0\%} = 14.4\%$$

The final change to "de-bottleneck" amine circulation is shown in Figure 5-1. Amine used to scrub the fuel gas has to be regenerated to a low residual acid gas loading. That is, fuel gas needs to be scrubbed down to 100 ppm of $H_2S$. Amine used to scrub hydrogen recycle gas on a hydro-desulfurizer need only scrub recycle gas down to 1000 to 2000 ppm of $H_2S$. It may take an extra 10% of reboiler steam to regenerate the lean DEA for fuel gas scrubbing as compared to hydrogen recycle gas scrubbing. As 30% of the amine circulation was used for hydrogen recycle gas $H_2S$ absorption, I calculated that the potential increase in amine circulation that would result from this modification would be

$$30\% \times 10\% = 3\%$$

To effect this improvement, I designed the new 8-in. line shown in Figure 5-1. Valve A was closed and valve B was opened. Both low- and high-pressure pumps were existing. This modification is an adaptation of amine regenerators that produce both a lean and a semilean solvent. In this case the kettle reboiler is actually the bottom theoretical equilibrium state of the amine regenerator.

Increasing feed preheat by cleaning the rich amine-to-lean amine heat exchanger would reduce the reboiler duty. Reduction in the reboiler duty would also reduce the vapor load on the tower trays below the feed point. As the regenerator was limited by both tray flooding and heat input, an increased feed preheat of 10°F would increase the amine circulation capacity by 8%. The 8% figure includes the benefit calculated for reinsulating bare sections of pipe and replacing the removable tower manways, reboiler channel head, and bell head insulating covers not replaced after last year's turnaround.

To summarize the benefits to the amine circulation rate:

1. *Removing heat-stable salts:* 8.1%
2. *Reducing $CO_2$ absorption:* 3.6%
3. *Increasing DEA concentration:* 14.4%
4. *Producing a partly stripped lean amine for scrubbing $H_2$ recycle gas:* 3.0%
5. *Increasing feed preheat:* 7.0%
6. *Insulating integrity:* 1.0%
   *Total:* 37.1%

What do all these techniques have in common?

- No new technology is employed.
- No new process equipment is required.
- No engineering contractors or permitting authorities have to be involved.
- The work can be done piecemeal as opportunity arises, starting immediately.

But this approach to a project creates a moral dilemma—a contradiction in my two objectives on a project:

1. Do the best job I can for my client by minimizing material and construction costs while achieving the most cost-effective portions of the project.
2. Make a lot of money for myself.

My clients, who are mainly petroleum refineries, have an expectation as to how much a process design package and engineering charges should cost. Engineering charges, they expect, ought not to exceed 10 to 15% of a project's total cost, and the process design package should be less than 10 to 20% of the total engineering charge. Josh was not going to pay me $135,000 for my ideas but only $13,500 for my time.

## SULFUR PLANT CAPACITY EXPANSION

One of the principles of my new religion is my perception of time. The past is dead and gone. The future is unknown and unknowable. But the present belongs to us. What can I do today, or tomorrow, or this autumn to increase the capacity of the Shalom refinery sulfur recovery unit (SRU) in Texas? A project that requires bids and contracts and project planning meetings and environmental impact reports has no meaning to me. It's like promising a meaty bone to a dog tomorrow.

But my clients and employers have often become angry with me because I see time differently than others do. Someday, switch grass grown on salt marshes in Louisiana may be converted to biodiesel. Someday, someone may devise a method to sequester $CO_2$ emitted from coal-fired power plants in China. Someday, Suncrude in Alberta will learn to mine tar sands without releasing methane into the atmosphere. Poseidon Refining Co. has the "human element" and B.L.O.P. is going green. But all I have is today; and unless we work with what we have today, the problem of environmental change will continue to spin out of control.

What's to be done? Let's start at the Shalom refinery SRU in Texas. Hydrogen sulfide is converted to liquid, yellow elemental sulfur in accordance with the following:

$$H_2S + \tfrac{1}{2}O_2 + 2N_2 \rightarrow H_2O + S_{(l)} + 2N_2$$

This means that two-thirds of the vapor flow through the sulfur plant is nitrogen left over from the partial oxidation of $H_2S$. However, the feed to the SRU also contains three other components in addition to $H_2S$:

- Water vapor
- $CO_2$
- Hydrocarbon vapors

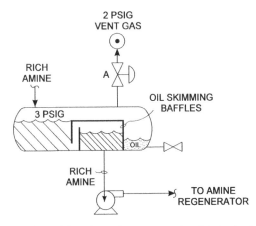

**Figure 5-2**   *Flash rich amine at low pressure to minimize hydrocarbons.*

The most common species of hydrocarbon (for a refinery that has a fluid catalytic cracker) in SRU feed is propylene. That is because light olefins are about three times as soluble in an amine solution as are paraffinic molecules. The oxidation of propylene in air yields

$$C_3H_6 + 4\tfrac{1}{2}O_2 + 18N_2 \rightarrow 3CO_2 + 3H_2O + 18N_2$$

This looks rather bad. Oxidizing 1 mol of propylene requires nine times as much air as does oxidizing 1 mol of $H_2S$. At the Mercury refinery in Stanlow, England, I greatly reduced the hydrocarbon content of the SRU's acid gas feed by reducing the flash drum pressure shown in Figure 5-2. For example, let's assume that the hydrocarbon content of acid gas is 2.0 mol% and that I drop the flash drum pressure from 35 psig to 10 psig. Then the new hydrocarbon content of the acid gas will be

$$2\% \times \frac{10 + 14.7}{35 + 14.7} = 1\%$$

Having reduced the hydrocarbon content of the acid gas feed by 1 mol%, I would multiply this 1% by a factor of 9 (because 1 mol of $C_3H_6$ consumes nine times as much air as does 1 mol of $H_2S$) to calculate a capacity increase of 9%. Note that at the Shalom refinery in Texas, I had to install a larger impeller in the pump shown in Figure 5-2 to accommodate the lower pump suction pressure. However, the NPSH (net positive suction head) available to the pump did not change because the vapor pressure of the rich amine had also dropped.

I have already discussed how to reduce the $CO_2$ in the acid gas from 20 mol% to 14 mol%. The $CO_2$ does not require any air and passes through the sulfur plant as an inert, except that it does increase $\Delta P$ through the SRU's vessels. Thus, a reduction of 6 mol% $CO_2$ will increase the SRU capacity by only 2%.

The same logic applies to water vapor in SRU feed. Minimizing the overhead receiver temperature at the regenerator can reduce the water vapor content from 7% to 2% and achieve another 2% in sulfur plant capacity. The only problem is that the acid gas feed line running to the SRU will salt up. Periodic water washing of this line to remove the water-soluble ammonia salts will then be required if the line is to run cooler.

In 1977, I developed the concept (but neglected to file for a patent) for the use of oxygen at a sulfur recovery plant at the Amoco Refinery in Whiting, Indiana. Also, I used up to 29% oxygen in enriched air at the Ever Hopeful refinery in Norco, Louisiana in 1981 with excellent results. The potential for increasing the SRU capacity using oxygen is now common practice, but is not practical at the Shalom plant in Texas, which was not connected to a source of pipeline oxygen.

Use of oxygen instead of air in sulfur plants may seem energy inefficient and hence environmentally evil. But this is not necessarily so, for several reasons:

- The air itself must be compressed to about 15 psig.
- The nitrogen in the air reduces conversion in the sulfur train.
- All the sulfur plant tail gas must be incinerated, and most of the tail gas is nitrogen.
- Nitrogen reduces the amount of steam generation in the high-pressure waste heat boiler (i.e., the main reaction furnace).

One thing we could do to increase capacity at Shalom in Texas was reduce the pressure drop in the first fixed-bed catalytic reactor. This reactor contains extruded catalyst ($\frac{1}{2}$ in. $\times$ $\frac{1}{2}$ in.) that partially plugs up with both sulfur and carbon deposits. To reduce this pressure drop, one runs the reactor at an elevated temperature for awhile. Usually, this reactor has the single largest pressure drop among all the reactors and condensers. The maximum reactor temperature is usually limited by the design temperature of the catalyst support screen.

I've saved the best for last—the single innovation in my 45-year career for which I am most proud: the cascaded seal leg. I introduced this design in 1977, also at the Amoco refinery in Whiting (see Figure 5-3) [1]. The idea is to keep the seal legs from blowing out. Liquid sulfur drains from the condensers through a U-shaped pipe 10 to 20 ft deep. If the pressure at the front end of the sulfur train gets too high, the first seal leg blows out and deadly $H_2S$ is exhausted to the atmosphere. By connecting the outlet from the first-stage condenser (shown in Figure 5-3) to the inlet of the second-stage condenser, the pressure at which the first seal leg would blow out was increased in proportion to the pressure in the second-stage condenser.

At Shalom, the SRU was indeed limited by seal leg blow-out pressure. The air blower had sufficient head and capacity for all potential acid gas flows as long as the air pressure could not overpressure the sulfur seal leg drains. Piping spool pieces were prefabricated and installed in a short shutdown. When combined with

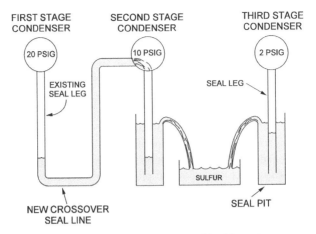

**Figure 5-3** *Piping change doubles seal leg blowout pressure.*

the other changes I have enumerated, the amine and sulfur recovery bottlenecks were eliminated.

Josh and I then went running together. This was my chance to race an Olympic athlete. We then had a few too many drinks, and he invited me home to meet his wife.

"Norman, my friend! You should also convert my wife to your new religion. She can lead the women's auxiliary," Josh said as we staggered up his driveway.

But Josh's wife proved to be a narrowminded person. She did not appreciate two drunken fanatics appearing in her parlor after midnight. Josh also has not been the asset I had hoped for. Having contributed $13,500 to the New Orleans chapter fund, he has yet to remit his annual contribution.

## SULFUR RECOVERY FROM SOUR WATER STRIPPER OFF-GAS

In an environmentally friendly process plant, sour water is stripped and then reused as wash water. The stripper overhead vapors consist of:

- $NH_3$
- $H_2S$
- $H_2O$

The $NH_3$ and $H_2S$ ought to be present in an equal molar ratio. The water content depends on the stripper reflux drum temperature. If this temperature is reduced too low (to minimize the moisture content of the off-gas), the line to the sulfur plant will plug with ammonia salts. This moisture (about 20 mol% of the sour water stripper off-gas) will reduce the temperature in the front end of the sulfur plant's main reaction furnace (Figure 5-4).

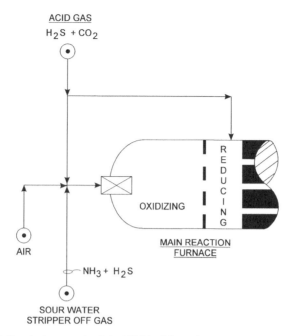

ACID GAS

$H_2S + CO_2$

AIR

$NH_3 + H_2S$

OXIDIZING

REDUCING

MAIN REACTION FURNACE

SOUR WATER STRIPPER OFF GAS

**Figure 5-4** *Sulfur plant conversion is lost if $NH_2$-rich gas is not combusted at high temperature.*

It's extremely important that this front compartment temperature be hot enough to decompose $NH_3$ to $N_2$. If the temperature is not about 2300 to 2400°F, or even hotter, the resulting nitric oxide salt accumulates downstream in the first catalyst bed. My experience is that conversion declines in the short run and builds $\Delta P$ over a longer period of time.

The conventional solution to this problem is to burn part of the acid gas, using 100% of the combustion air to generate the desired front-end oxidizing compartment temperature of the main reaction furnace. Then, as shown in Figure 5-4, the remainder of the acid gas is introduced into the back-end reducing compartment of the main reaction furnace, where the temperature is lower, before the mixture flows into the high-pressure steam boiler tubes.

The problem at the Himmler refinery in New Jersey was that the $CO_2$ content of the acid gas from the amine regenerator was not 10 to 20%, which is normal, but 70% $CO_2$ and 30% $H_2S$ (on a dry basis). The refinery had not designed their amine absorber to minimize $CO_2$ recovery. The fluid catalytic cracker (FCU) feed was low in sulfur. Also, the high FCU catalyst circulation rate had entrained a lot of $CO_2$-rich flue gas back into the FCU fractionator. As a result of these factors, the acid gas feed to the sulfur plant had 70% $CO_2$ and 30% $H_2S$.

On the other hand, the FCU feed was high in nitrogen, which produced lots of ammonia, which resulted in lots of ammonia in the sour water stripper overhead. The feed to the sulfur plant was therefore about 30% sour water stripper off-gas. As the destruction of $NH_3$ to $N_2$ is an endothermic reaction (heat is absorbed), large amounts of heat have to be liberated by oxidizing $H_2S$ to $SO_2$ in the oxidizing compartment of the main reaction furnace. The problem is that when $CO_2$-rich acid gas is charged to a sulfur plant, even without any sour water stripper off-gas, it becomes difficult to keep the reaction furnace hot enough. Supplying the extra heat for the sour water stripper off-gas proved to be impossible. As a result, the oxidizing section of the reaction furnace was run too cold to decompose the $NH_3$ efficiently.

Then, to add to this sad story, the interstage reheat was not provided by indirect 600-psig steam reheaters, or through bypassing part of the hot, main reaction furnace effluent to the downstream reactors. Rather, the reheat was provided by combustion of acid gas. This further reduced heat to the oxidizing section of the main reaction furnace.

The result of these design errors, plus ambient heat losses and moisture in both feed gas streams, made it quite impossible to run the oxidizing, front-end section hot enough to decompose the $NH_3$. The relatively high hydrocarbon content of both feed gas streams (over 1%), partly mitigated the heat balance problems, but unfortunately, promoted coke formation in the downstream catalyst bed.

## RESULT OF LOW REACTION FURNACE TEMPERATURE

So the oxidizing section operated several hundred degrees Fahrenheit below that required for $NH_3$ destruction, even with the excessive hydrocarbon contamination of the feed. As a result:

- Overall $H_2S$ conversion to sulfur was only 70% (a good three-stage sulfur plant can achieve 97%).
- The $\Delta P$ value across the first catalyst bed rose after a few months to 2 psi (normal is 0.5 psi).
- The sulfur product was brown (every sulfur plant that I have worked on produced bright yellow sulfur).

As an indication of the very low conversion of $H_2S$ to sulfur, the temperature increase across the first catalyst bed was only 60°F. It's supposed to be 150°F. Also, the temperature rise across the third (final) catalyst bed was 25°F. It's supposed to be zero, because the conversion reactions should be 99% completed in the upstream reactors.

That the low sulfur recovery was due mainly to the $NH_3$-rich gas was proven by stopping its flow to the sulfur plant. Conversion rose from 70% to 90%. It appeared that the nitric oxide salts that had escaped into the first catalyst bed had caused a loss in catalyst activity. This is not normally observed on plants that operate with a proper main reaction furnace oxidizer section temperature above 2300°F when processing sour water stripper off-gas.

## FRACTIONATION BETWEEN NH$_3$ AND H$_2$S

My client, Himmler Oil, planned to rectify this problem by constructing two towers to reprocess the sour water:

- One tower would distill the H$_2$S overhead and leave the NH$_3$ in the bottoms.
- A second new tower would distill the NH$_3$ overhead and leave the clean water in the bottoms.

The H$_2$S would flow to the sulfur plant. The NH$_3$ could be sold as a fertilizer.

This process was introduced in the 1960s. I have seen it not working too well at several locations, and not working at all at others. I guess it must work somewhere, but not in the 10 or so locations that I've visited. But one thing is certain. This is a complex, expensive undertaking requiring two new towers, one of which operates at high pressure.

Having a divine mission to avoid new process plant construction, I proposed an alternative plan to the Himmler refinery.

## USE OF OXYGEN TO PROCESS NH$_3$-RICH GAS

When I first used oxygen at the Ever Hopeful refinery in 1981 to increase sulfur plant capacity, I noted an increase in the main reaction furnace temperature of 150 to 200°F. The reason for this is that there is less nitrogen to absorb the heat of combustion of H$_2$S. Each mole of oxygen used eliminates 4 mol of nitrogen. Each mole of oxygen used thus cancels out the same amount of heat absorbed by about $2\frac{1}{2}$ mol of CO$_2$. Thus, oxygen-enriched air permits acid gas with a high CO$_2$ content to be used to decompose NH$_3$ in the sour water stripper off-gas.

As Himmler Oil was already using oxygen-enriched air in their FCU, the addition of oxygen to the sulfur plant combustion air was relatively simple. (See reference [1], which I published before actually using oxygen at the Ever Hopeful sulfur plant in Louisiana in 1981.) Clearly, this was a more environmentally friendly way of dealing with Himmler Oil's problem than building a new process plant to fractionate H$_2$S and NH$_3$ would be.

I'll make two additional observations pertaining to using oxygen in sulfur plants:

- Be careful to use oil-free piping and to keep the velocities low. Otherwise, the steel line may catch fire.
- The higher the temperature in the main reaction furnace, the more the H$_2$S decomposes to hydrogen and elemental sulfur. This further reduces the need for combustion air, with further enhancement in the sulfur plant capacity and conversion efficiency.

## REFERENCES

1. Lieberman, N. P. "Effective Sulfur Plant Operations," *Oil and Gas Journal,* Mar. 17, 1980, p. 155.
2. Kohl, A., and Risenfeld, F. *Gas Purification*, 2nd ed., Gulf Publications, Houston, TX, 1974.

# Chapter 6

# *Treating and Drying Hydrocarbons*

Bad news! I have offended the religious community of New Orleans. I had been invited to address other leaders at an important convocation. The subject of the gathering was: "The Power of Prayer in Rebuilding New Orleans after Hurricanes Katrina and Rita."

My presentation was not well received. "Dear fellow clergymen. The message is clear: Human beings should not build houses 12 ft below sea level. We should not be pumping rain back into the Gulf of Mexico. It wastes too much energy. Better to abandon New Orleans and move to higher ground and not resist the forces of nature."

Rabbi Silverstein followed me to the podium. "On behalf of the entire Hebrew community of Orleans Parish, I would like to apologize for Mr. Lieberman. I have known Mr. Lieberman for many years, and his rude remarks are consistent with his past behavior."

Fortunately, not everyone was quite as hostile as Rabbi Silverstein. Robert Glass, the manager of Jupiter Oil refinery, rushed over to congratulate me on my presentation.

"Wonderful, Mr. Lieberman. Just wonderful!" Mr. Glass exclaimed. "Your ideas are so practical, yet full of hope and inspiration. I feel uplifted, thanks to your remarks!"

"Really?" I asked. "You seemed somewhat angry over that crude pre-flash tower project."

"No, Norm. I'm extremely pleased. The pre-flash naphtha is water white. Perfect reformer feed. I am injecting a few ppm of a silicon-based defoaming chemical into

*Process Engineering for a Small Planet: How to Reuse, Re-Purpose, and Retrofit Existing Process Equipment,* By Norman P. Lieberman
Copyright © 2010 John Wiley & Sons, Inc.

the pre-flash tower's feed to suppress the foam formation in the tower's bottoms. Also, I've installed a conductivity probe on the overhead reflux that sounds an alarm when we get water in the reflux due to high interface level in the water draw-off boot. Better than building a new pre-flash tower," Mr. Glass winked knowingly.

"But what are you doing here, anyway?" I inquired.

"Oh, I'm affiliated with a local faith-based congregation. But listen, my running into you like this is a happy coincidence. I have a project that requires a quickie process design. Do you know how to design a salt dryer for jet fuel? We have a haze problem."

"Sure," I answered. "What's the feed rate?"

"Size it for 15,000 bsd. The existing salt dryer produces a hazy jet at anything over 8000 bsd. The problem is getting gradually worse. I'd love to have you on our project team, Norm. The process design package has to be on a cost plus basis with a limit of $135,000. That must include a vessel sketch for the new salt dryer and an orientation drawing of all new process nozzles.

"Mr. Glass, did you say $13,500 or $135,000 for the dryer vessel sketch?"

"Why, $135,000," Mr. Glass laughed. "Only one problem. I'll need a pro-forma invoice. The project is in this quarter's budget. So I'll have to pay you in advance. I hope that's not a problem?"

"Oh, I guess that will be okay."

## TROUBLESHOOTING A SALT DRYER

I could design the new salt dryer in one day. But perhaps there was a problem with the existing dryer that I should avoid in my new design. So I drove out to the plant to observe the current operation of the jet fuel treater, shown in Figure 6-1.

The jet fuel was first treated in a NaOH scrubber to extract organic acids such as naphthenic acid. The resulting sodium naphthanate is a salt of a weak acid and a strong base, which means that the dissolved salt would produce an alkaline pH. Indeed, when I checked the aqueous phase in the caustic scrubber, it was, as expected, pH 14.

I then checked the pH in the second vessel, the water wash. The water wash is critical. Some sodium naphthanates and caustic will always entrain from the naphthanic acid extraction caustic wash vessel. If the sodium compounds escape into the salt dryer, they will coat the salt in the dryer and make the salt less accessible to the jet fuel. It is the function of the water wash vessel to remove the entrained sodium naphthanates from the jet fuel before the jet fuel enters the salt dryer. If the salt ever becomes coated with the sodium naphthanates, a soapy type of substance, the water dissolved in the jet fuel will not be absorbed effectively by the salt. The result would be hazy jet fuel.

The pH of the water drained from the water wash (i.e., the middle vessel) was slightly above neutral, at pH 8. Evidently, I thought, not very much of the sodium naphthanate soapy substance was being carried out of the caustic wash and into the water wash. That is what I thought, but I was quite wrong.

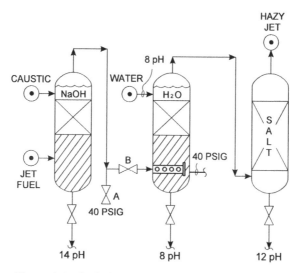

*Figure 6-1* Jet fuel treater with a contaminated salt dryer.

The final vessel was the salt dryer, the vessel that was to be replaced with my new design. The dryer had 20 ft of rock salt, of the sort used to deice winter roads. Water trickled from the salt dryer drain. This water had been carried out of the water wash vessel by the jet fuel. Logic indicated that its pH be 8 or less. However, my pH test strip indicated pH 12. How could the pH of the water draining from the salt dryer be higher than the pH in the water wash vessel?

pH is an exponential function. Water with a pH of 12 has a 10,000-fold higher concentration of alkaline molecules than that of water, with pH 8. The only possible explanation for this pH pattern is that the effluent from the caustic wash vessel was partly bypassing the water wash. As there was no external bypass pipe, the bypassing had to be inside the water wash vessel. Clearly, then, what I was observing was poor mixing between the aqueous and hydrocarbon phases in the second vessel, shown in Figure 6-1. More to the point, there was nothing wrong with the existing salt dryer.

## SALT DRYER INTERNAL CONDITION

I asked Craig Hebert, the lead operator on the unit, about the history of the hazy jet fuel. "Mr. Lieberman, haze in jet fuel is caused by dissolved moisture. The jet fuel becomes saturated with moisture when it's steam stripped. Steam stripping is required to meet the 110°F ASTM (American Society of Testing and Materials) flash point specification for aviation jet fuel. Unless the majority of this dissolved water is extracted by the salt, the jet fuel becomes hazy due to water precipitation when the jet fuel cools below 90°F. When we started this unit 10 years ago, we could make haze-free jet fuel at 15,000 bsd. However, after a few years we began to have trouble. To meet the haze point spec, we had to reduce the feed rate. I thought the salt had been washed out of the dryer. But when we opened the salt dryer, I was surprised. The dryer was still full of salt. The salt had settled down a bit, and it looked somewhat dirty."

"Craig, did the salt have channels cut through it?"

"No, Mr. Lieberman, it was okay. We just topped it up with a few bags of rock salt and started back up. You know what I think? That something has gotten on the salt and ruined it. I mentioned this to Mr. Glass, our boss, but he just became angry."

"Really," I asked, "Why?"

"I guess we did something wrong that ruined the salt. I guess dirty salt doesn't dry."

"No, Craig. That's not what I asked. Why did Mr. Glass get angry?"

"Mr. Lieberman, I told you. It's like I said," answered Craig. "It's because we ruined the salt. We did something wrong. I don't know what. We screwed up that salt dryer. Now we got 700,000 barrels of hazy, off-spec jet in the tank field. It's our fault. That's why Mr. Glass is mad. Do you have any idea how we can fix it up to get rid of that haze?"

"I sure do," I replied.

"Yeah? The new salt dryer vessel. I guess that's the ticket?" Craig said in a puzzled voice. "But you know that the salt dryer used to work pretty good. But that was years ago."

## FIXING THE WATER WASH

The heat and humidity pressed down on Chalmette, Louisiana. It was only May. What would August be like? This was hotter than May 2005, the spring before Katrina and Rita. I had lived in New Orleans for 30 years. Last season was the first year I had a great crop of bananas and papayas. Global warming was turning my backyard into a tropical paradise. The last hard freeze was in 1996. No wonder, my tropical fruit trees were thriving.

"Let's look at the water wash vessel. Grab your pipe wrench," I said as I picked up my gloves. "Craig, connect up that 1-in. water hose to valve A (shown in Figure 6-1).

"Mr. Lieberman, shouldn't I check this out with the shift supervisor first?"

"Uh, no. Don't bother. I've already checked with Robert Glass. He's approved trying this. It's okay to call him, extension 400," I said, handing Craig my cell phone. "He's at a meeting with the V.P."

"What water rate are we looking for," Craig asked.

"Whatever water rate flows through a 1-in. hose," I replied. (From the designer's perspective, I would use about 1 lb of the aqueous phase for 10 lb of hydrocarbon. At the required feed rate of 10,000 to 15,000 bsd, that is about 30 gpm of water. This is just a little more then would typically pass through 100 ft of 1-in. hose with a 20-psi $\Delta P$.)

After Craig connected the water hose to the hydrocarbon inlet on the water wash vessel, I throttled valve B. I was going to use B as a mix valve to bring the contaminated jet fuel into intimate contact with the clean water. It would have been best to use an inline static mixer or, better yet, an adjustable globe valve rather than an ordinary gate valve. Still, an isolation valve can be used for mixing if the gate is closed sufficiently to generate a pressure drop of 15 to 25 psi. But the real advantage of the gate valve was that it was the only valve there.

I closed valve B until the stem protruded from the hand wheel by 2 in.

"What's this all about, Mr. Lieberman? What are you trying to accomplish?"

"Craig, you didn't ruin the salt. It's a mechanical failure of the jet fuel pipe distributor in the bottom of the water wash vessel. I measured the $\Delta P$ across the pipe distributor; there was none. But the calculated pressure drop was 12 psig. I think the distributor must be damaged."

"No, Mr. Lieberman. It's not damaged. It's not there. I remember when we first loaded the salt. There is a 4-in. internal flange inside the vessel, but there's nothing bolted to it. I guess your pipe distributor was never installed. I guess that the jet fuel just channeled up one corner of the water wash vessel. It never came into contact with the water. The soapy stuff—what did you call it?: sodium naphthanates—got carried over into the salt dryer. How else could the pH be higher in the salt drum than in the water wash drum?"

"Yes, you're right, Craig."

"Yeah! And then," he continued, "the salt got covered with the sodium stuff. Probably took a few years. So the salt and the jet fuel don't come into good contact. So the water dissolved in the jet mostly slips out of the dryer. Then, hazy jet."

"Right, Craig," I said, "So now we will mix the jet fuel and water through the mix valve. It will take the place of the pipe distributor. The wet jet fuel will, I hope, carry over into the salt dryer and wash the sodium naphthanate off the salt. It will probably take some time, maybe a few days or weeks."

"But, Norm," asked Craig, "How can we tell if it's working? If we don't have a test, how do I and the other shifts know if we're doing any good?"

"Look, Craig, if the pH on the wash water drain goes up and the pH on the salt dryer drain heads down, it's positive proof we're going in the right. . . ."

"Because we will be doing a better job of extraction of the sodium naphthanates in the wash water and preventing contamination of the salt," Craig concluded.

A spark of common purpose and understanding passed between us. A bond of mutual respect had been forged. Better to light one small candle than curse the darkness.

Four days later, the pH levels at both the water wash and salt dryer had lined-out at about 10 pH. At rates above 12,000 bsd, the jet fuel product was haze-free. I wrote a longish report about my field work to Jupiter Oil, with the conclusion that the new salt dryer vessel could not be justified, as the problem of the hazy jet fuel product was gone. Based on achieving the objective of the project—haze-free jet fuel—and the fact that I had been paid in advance for that purpose, I felt that the contract conditions had been fulfilled. No vessel sketch would be required.

The next day I received a notarized letter from the Jupiter Oil's legal department. I was to return their check at once. I was in gross breach of contract and no compensation could be expected for engineering time expended on this project. Check number 666, in the amount of $135,000, was to be returned to Jupiter Oil immediately, to avoid further legal action.

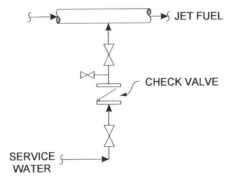

**Figure 6-2** *Check valve required when connecting a utility to a process line.*

To say that Mr. Glass was angry would be a gross understatement. It would be best if I reproduce his email in its unedited version:

To:   Mr. N.P. Lieberman
      Chemical Engineer
      New Orleans, Louisiana

I am terminating permanently and irrevocably all contracts between you and Jupiter Oil. You have initiated a major process revision at our Merox jet fuel treater without recourse to a hazop review or JAP (Jupiter-approval procedure) or even MOC (management of change). Furthermore, you have recklessly caused a utility system (service water) to be connected directly to a hydrocarbon process line (jet fuel) without an intervening check valve to prevent reverse flow of a hydrocarbon into service water [see Figure 6-2]. Jupiter Oil has, at substantial expense, been forced by your careless engineering to replace your hose with 120 ft of screwed $1\frac{1}{2}$-in. piping, the appropriate check valve, and a double block and bleeder.

Finally, your actions have marginalized the economic basis for the new salt dryer vessel, which had already been approved by corporate management, with $1.8 million appropriated.

Robert I. Glass, Sr.
Plant Manager

P.S.   I am canceling your invitation to our November 2 "Day of the Dead" coven in the French Quarter, as per Diane's request. R.I.G.

## AN APOLOGY

I made up the part about Jupiter Oil. The actual incident occurred in 1987 at the Mars refinery in England. It was worse in the sense that Mars Oil had installed, at great expense, a large boiler feed water injection pump to increase the flow of makeup wash water, all of which was wasted: as indicated by the pH of the makeup water and the wash water drain (see Figure 6-1) being identical. The expense of the treating chemicals and the energy to prepare the boiler feedwater, used on a once-through basis, was completely wasted, as was the new pump.

The pipe distributor for the jet fuel in the wash water vessel had been installed. But both the calculated pressure drop through the orifice distribution holes and the measured pressure drop was zero. When I phoned the VOP engineering contractor to ask about their design, they told me that their procedures were confidential. My design of a pipe distributor is, however, not confidential.

## PIPE DISTRIBUTOR DESIGN

- Calculate the velocity in each arm of the distributor pipe.
- Size the distribution holes so that the hole velocity is not less than three times the velocity in the arms.
- This procedure will ensure that the controlling pressure drop will be through the distribution holes, not through the pipe arms.
- Orientate the holes 30° above the horizontal plane, equally distributed.
- The smaller the holes, the more holes and the better the distribution, but the more likely the holes will plug. Ordinarily, a $\frac{1}{4}$-in. hole is a good compromise between plugging and distribution.
- A $\frac{1}{2}$-in. or larger drain hole is essential at the low point of the distributor. Safety requirements demand that all portions of the vessel internals be self-draining to aid in hydrocarbon freeing vessels during unit turnarounds.
- Do not support the distributor arms rigidly on the vessel walls. The arms must be provided with latitude to move because of thermal expansion.
- Never construct distributors out of carbon steel pipe, as the high hole velocities are necessarily somewhat erosive.

The Mars refinery had lived with this problem for a decade by allowing haze to slowly settle out in jet fuel product tanks. The engineer at VOP argued with me about my distributor design criteria. Now I wouldn't waste time arguing, I would just explain that I am a Messenger of Zeus, appointed to avoid useless capital investments by preaching the gospel of proper pipe distributor design procedures rather than building useless new salt dryers, or wasting boiler feed wash water due to poor contracting efficiency, or attending "day of the dead" covens with witches and warlocks down on Bourbon Street.

## MERCAPTAN SWEETENING

In 1969, I cleverly managed to get myself demoted from senior process designer in the Amoco Chicago Engineering Division to the position of subassistant technical service plant engineer at the Whiting, Indiana refinery. But it wasn't my fault. I was just trying to follow my mother's advice to use what we already have because we live on a small, poor planet. Here's what happened.

The No. 4A treating plant had been built before I was born. It converted evil-smelling mercaptan sulfur in gasoline to odorless disulfides:

$$RSH + RSH + O_2 \rightarrow RSSR + H_2O$$

where
  R = hydrocarbon group
  SH = mercaptan sulfur
  RSSR = disulfide

The equation above is called the sweetening reaction, and the process itself is called gasoline–air sweetening. The No. 4A treating plant had a capacity of 24,000 bsd. It was originally designed at 10,000 bsd in 1940. As refinery rates increased in the 1960s, the required capacity for No. 4A approached 30,000 bsd. I was assigned to develop various case studies to replace the archaic sweetening plant with an up-to-date process, to achieve a capacity of 36,000 bsd. Every month I would present the results of my study to Dr. Horner, the vice president of engineering, at the project review meeting. And every month, Dr. Horner would reject my study:

- "Lieberman," Dr. Horner would say, "we need to use proven technology."
- "Lieberman, we need to come up with an innovative and creative design."
- "Lieberman, the licensing fee for this process is excessive."

I happened to mention Dr. Horner's hostile attitude to my mother once, and she said, "Listen, Norman! Use what you have."

"Mom. We can't. The plant's too small."

"So why is that?" my mother asked.

"I don't know, mom."

"So why don't you find out, Mr. Smarty Engineer?"

So I drove over to Whiting, Indiana and asked the panel board operator: "What's the process limit on the No. 4A naphtha sweetening plant?" (shown in Figure 6-3).

"It's 24,000 bsd," answered the old panel operator. "Our sour naphtha flow indication stops at 24,000 bsd. It's the maximum scale reading. Say, who are you anyway? You got permission to be on my unit?"

"Dr. Horner sent me," I explained. "But why can't you just raise the flow above 24,000 bsd, even though you can't measure that extra naphtha flow?"

"Who is Dr. Horner? You're not one of those Chicago engineers are you? I have to know my feed flow so I know how much air to use. I multiply the naphtha mercaptan concentration from the lab sample, times the naphtha feed flow and that sets the rate

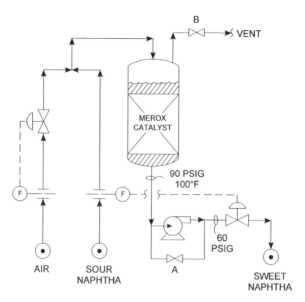

***Figure 6-3*** *Naphtha-air sweetening plant.*

of sweetening air I need. See, if I don't know my feed rate, I can't figure out how much air I need."

"Well yes," I said, "But why don't you just use lots of excess air? It's only plant air, and the air control valve is only 20% open?"

"Say, who are you? Are you a retard? You ain't one of those dumb headquarter enginners from Chicago? Listen Mr. Engineer, if I use more oxygen than is consumed in the sweetening reaction that converts mercaptans to disulfides, the vapor space at the top of my Merox reactor will get oxygen accumulating in it. Then I'll have an explosive mixture of hydrocarbon naphtha vapors and oxygen in the vapor space at the top of my reactor. I've got to crack open the vent valve B all the time, to purge out air accumulating in that vapor space."

"Explosive mixture? How can anything explode in the reactor?" I asked. "There's no source of ignition in the Merox reactor, is there?"

"Man! You people in headquarters aren't just stupid, you're dangerous! Ever hear of static electricity? Ever hear of pyrophoric iron [$Fe(HS)_2$]? We blew up a propane vessel in the refinery 10 years ago by not being careful about using too much excess $O_2$."

"Okay. I agree. You're right. But how about reranging the naphtha feed flow meter," I suggested.

The unshaven, gray-haired operator glanced up at the wall clock. "What flow you want?"

"How about 36,000 bsd maximum?"

"Javorski," he screamed into the control room kitchen, "Get off your fat butt. I need you to change the range tube on FRC-36 from 100 in. water to 225 in. $H_2O$. Pronto!"

(*Note:* $\Delta P$ varies with flow squared. If I want to increase the maximum flow I can read by 50%, that is, from 24,000 bsd to 36,000 bsd, the new orifice plate $\Delta P$ will increase by a factor of $1.5^2$, or 2.25. The old maximum $\Delta P$ that could be measured was 100 in. $H_2O$ of water, and $2.25 \times 100$ in. $H_2O$ is 225 in. $H_2O$.)

Twenty minutes later, the sour naphtha flow meter was reading only 67% of the chart scale instead of 100%, but still at the same 24,000 bsd.

"Guess you'll want me to raise up the naphtha feed and sweetening airflow now?" the panel man asked in a less aggressive tone. But when he opened the naphtha control valve shown in Figure 6-3 to 100%, the flow came up to only 29,000 bsd.

"Sorry, but I need 30,000 bsd."

"Yeah, and I need me a million bucks. You go figure out what to do. The FRC's wide open and my relief will be here soon."

So I went onto the unit and looked at the product booster pump. The pump discharge pressure was 30 psig lower then the pump suction pressure. The brave little pump had been designed in 1940, before I was born, for 12,000 bsd. It was so far above its design flow that it was restricting, rather than boosting, the flow at 29,000 bsd.

Maybe I should open valve A (Figure 6-3), I thought. It seemed crazy to open a bypass around an operating pump to get more flow. But the idea of a pump choking the flow was also crazy. However, it wasn't the concept of opening the pump bypass valve that was actually bothering me. The big problem was that engineers from Chicago were not allowed to manipulate refinery process valves on operating units. At least not without getting an okay from the panel operator. And I was pretty certain that this operator didn't like me.

So I waited around until 4:00 P.M. And just at shift-change time, I spun valve A open. When I went back into the control room at 4:30 P.M., a tired-looking middle-aged lady was sitting in front of the panel.

"Excuse me ma'am," I said, "But I don't believe you're using enough sweetening air. The naphtha FRC's reading 32,000 bsd. The chief operator on the previous shift raised the naphtha flow from 29,000 bsd to 32,000 bsd, but I guess he didn't have time to raise the airflow, too."

"Okay. Sure. I'll do it," she said, "But who are you, anyway?"

"I'm the manager of the treating technology division in the Chicago engineering headquarters center. Dr. Horner and I are here for a planning meeting with Dr. Swearinger, the refinery manager. You know Dr. Swearinger, don't you?"

At the next project meeting with Dr. Horner, I reported on my progress:

- The capacity of the No. 4A treating plant now exceeded requirements, with a demonstrated capacity of 32,000 bsd.
- All naphtha production was "doctor" sweet and on spec for mercaptan content.
- I had closed out the project file, and no further case studies would be forthcoming for new naphtha sweetening facilities at the Whiting refinery, as the existing plant had adequate capacity.

The Vice President glared at me with naked hatred. "Lieberman, who authorized you to open process valves on an operating unit? The plant operators' union at the Whiting refinery has lodged a formal complaint about your reckless actions and aggressive attitudes."

So that's how I came to be demoted to subassistant engineer at the No. 12 pipe still crude unit in Whiting, Indiana, in 1969.

# Chapter 7

# Minimizing Process Water Consumption

I should apologize not only to Jupiter Oil for my fabrications about their jet fuel drying problem, but also to Mercury Oil in Washington. My threat to liquidate their plant manager was entirely inappropriate. Mrs. Hightower has every right to express her confidence in that refinery's future, free of threats from religious fanatics such as myself.

My students at the troubleshooting seminar I presented at the Mercury refinery in Washington state should not have invited me to the refinery's fiftieth-year anniversary dinner. It's their fault, not mine. They probably knew that Mrs. Hightower would say, "We have completed 50 years of successful refinery operations, and with good fortune our new central control room will usher in another 50 years of safe and efficient operations." My students, especially Charlie, knew perfectly well the irrational response I would have to Mrs. Hightower's provocative speech.

First, we certainly cannot continue to run our crude oil refineries in the future as we have in the past. It's true, we have made the same products, to the same specs, using the same equipment for the last half century. I know. I completed my first retrofit design in 1965 to the No. 12 pipe still at the Amoco Refinery in Whiting, Indiana.

Possibly, we have enough crude oil (virgin and synthetic) to continue charging 100 million bsd through our refineries for another 50 years. There is actually a lot of crude left in Russia, Iraq, Venezuela, Kazakhstan, Iran, and all those other anti-American countries. Political developments are not subject to engineering calculations. Not so $CO_2$ accumulations in the atmosphere and the oceans. The effect

*Process Engineering for a Small Planet: How to Reuse, Re-Purpose, and Retrofit Existing Process Equipment,* By Norman P. Lieberman

of heat-trapping gases ($CO_2$, methane, NO$x$, flurocarbons) is subject to calculation if one takes into account the effect of $SO_3$ emissions and particulates.

I had thought my design work on expanding sulfur recovery and amine circulation capacity (Chapter 5) was a positive contribution to the planet, to reduce the acid rain formed when $SO_3$ reacts with moisture. But what I had failed to understand is that sulfates ($SO_4^=$) reflect sunlight, as do particulates, and thus temporarily diminish global warming. Sulfates and particulates cause solar dimming, which has temporarily slowed global warming in the last half of the twentieth century by about 0.5°F. Thus, the refining industry's huge and successful efforts to improve conversion of $H_2S$ to liquid sulfur in the past few decades has accelerated heat accumulation on the planet. Again, the effect of sulfates on global warming is temporary. Overall, surface temperatures have increased by 1.3°F from 1907 to 2009.

With reduced sulfate and particulate emissions, the accumulation of heat in the oceans will make human activities quite different before 50 years has passed. For instance, the Mercury refinery will be under water. This is not speculation, simply extrapolation. Not to say that the Mercury refinery will not be submerged before 2035. That is quite possible, too. But if Mercury and all the world's other refineries, coal-fired power plants, ammonia fertilizer production facilities, and petrochemical plants keep perking along at current rates of about 220,000,000 fuel oil equivalent barrels (FOED) per day, then by 2058, Mercury's new control center will definitely be submerged.

## TWO-STAGE WASTEWATER STRIPPER

Second, what really unhinged me was not fuel, but water. Not the rising waters of Puget Sound, but the wasteful use of process water in the refinery. For example, the crude unit desalter at the Mercury plant is using once-through fresh wash water. Most refineries use stripped sour water as desalter wash water. For a 150,000-bsd refinery, this reduces the plant effluent about 300 gpm. I asked Chris, the crude unit engineer, why they don't use stripped sour water. Chris responded that the stripper bottoms had excessive $NH_3$ (100 ppm). If $NH_3$ is not stripped down to 10 to 15 ppm $NH_3$, Chris correctly noted that excessive chlorides are left in the crude charge. The chlorides hydrolyze to HCl, which causes corrosion in the crude fractionator overhead condensers. Also, the tower trays above the jet fuel draw-off will plug with corrosion products from the reflux drum.

To overcome the problem of excessive $NH_3$ in the sour water stripper bottoms, the stripping steam rate should be increased. But if the extra vapor flow caused the trays to flood, more stripping steam will make the $NH_3$ concentration increase rather than decrease. And tray flooding was the limiting factor at the Mercury sour water stripper, constraining their use of stripping steam.

There is an inexpensive retrofit to overcome this limitation. As shown in Figure 7-1, a new side draw-off is installed at tray 15. This partially stripped water is then used as wash water in the hydro-desulfurizer. The hydro-desulfurizer wash water is normally purged at $NH_3$ concentrations in excess of 10,000 ppm. Thus, if

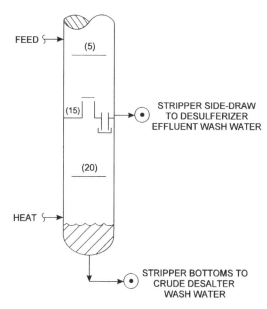

**Figure 7-1**  *Two-stage sour water stripper.*

the makeup water contains a few hundred ppm of $NH_3$, it increases the purge water rate by only a few percent.

If half of the water is extracted at tray 15, the $L/V$ ratio (i.e., the stripping factor) on trays 16 to 20 will double, which will drop the $NH_3$ content of the stripper bottoms by an order of magnitude. In summary, I have created a two-stage sour water stripper.

The first stage produces poorly stripped water to extract $NH_3$ salts from the hydrosulfurizer reactor effluent. The second stage produces well-stripped water to remove NaCl, $CaCl_2$, and especially $MgCl_2$, from crude charge. This two-stage system will require a separate low head pump for the stripper bottoms. No additional cooling is required, as desalter wash water is best kept hot so as to increase crude preheat.

## STEAM CONDENSATE RECOVERY

I have written a long email to Mrs. Hightower, but not only about the desalter wash water. I also included a design to stop the loss of steam condensate to the sewer. But I fear my email has been caught by a spam filter; I never heard a word back from Mrs. Hightower. So I will repeat my advice here. Maybe she will be divinely inspired to buy my book. Who knows? The universe is a mysterious place.

Steam condensate gets dumped to the sewer for two reasons:

1. Steam or water hammer
2. Condensate drum–level control problems

Both of these difficulties originate because saturated water boils when it undergoes a reduction in pressure. For example, 328°F and 100 psig saturated water depressured to 25 psig and 268°F will generate 3 wt% steam. The 3% doesn't seem like a big number, but it is. The volumetric flow increases by about 1700%. So if pipeline velocities are designed at 5 ft/sec for water, the line velocity will accelerate to 85 ft/sec for the steam trap effluent. This creates some nasty problems.

The water, carried down the pipe by the speeding steam, will at some point change direction at an elbow or tee-junction fitting. As a high-velocity slug of water contacts a 90° elbow, it bangs against the elbow: thus, steam hammer. This conception of the cause of steam hammer is somewhat a guess on my part. But what is definite is that steam hammer is a consequence of high-velocity flow of two phases inside a pipe.

Condensate drum-level control problems and the resulting water backup in the steam side of reboilers and steam heaters is also a consequence of high velocity in piping. As shown in Figure 7-2, the velocity in the 3-in. condensate drain line increases from 5 ft/sec to 85 ft/sec. The reduced density and increased velocity,

$$\Delta P = (\text{density}) \ (\text{velocity})^2$$

will cause the pressure drop per length of line to increase by a factor of 17. The greatly increased line loss will also cause water backup in both the condensate drum and the reboiler. The waterlogged tubes in the reboiler transfer only 1% of the heat as compared to the unsubmerged tubes. Condensate backup in the reboiler results in subcooling of the steam condensate. This suppresses steam evolution and $\Delta P$ in the 3-in. condensate drain line. In this way, condensate backup restores hydraulic

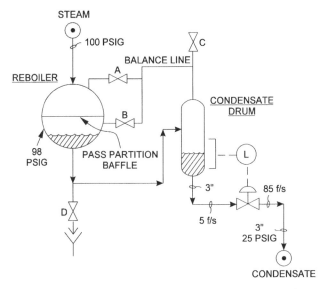

*Figure 7-2* Steam condensate drum will not drain properly.

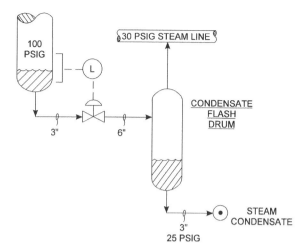

**Figure 7-3** *Condensate flash drum surpresses condensate backup in the process reboiler.*

equilibrium to the condensate drainage. The price we pay for condensate backup is then the loss of reboiler duty as the condensate is subcooled.

Regardless of the problem, steam hammer or loss of reboiler duty, plant operators have a standard solution. We all know the unfortunate answer. Valve D, shown in Figure 7-2, is opened. Valuable steam condensate is lost to the sewer. More of our little green planet's freshwater reserves are then drawn upon to make new BFW (boiler feedwater). What's to be done?

One possible solution is to size the level control valve on the condensate drum, and the downstream piping for the evolved steam. Perhaps the downstream piping should have been increased from 3 in. to 6 or 8 in. If the new 6-in. pipe has to go only a few hundred feet to a large condensate collection header pipe, I would use the 6-in. line option.

If the line has to travel a 1000 ft before it ties into a larger line, I would use an intermediate flash drum, as shown in Figure 7-3. The steam flow venting into the 30 psig steam header will be 3% of the steam supply to the reboiler. The new flash drum is just a section of large-diameter vertical pipe, sized as a steam–water separator. Depending on system hydraulics, a condensate recovery pump may also be required on the flash drum. That's getting expensive, but not as expensive as diverting steam condensate to the sewer for 50 years.

## CONDENSATE DRUM BALANCE LINE LOCATION

One of my problems in dealing with Mrs. Hightower, the plant manager, is that I'm a residual male chauvinist. But thanks to Sandy Lani, I am mostly cured. Sandy was killed in a motorcycle accident in 1985. Shortly before her tragic death, she explained a design error that I had been making for three decades.

Referring to Figure 7-2, note the location of the balance line. It can be connected to the channel head of the reboiler at valve A or valve B. I, and most other designers connected the balance line at the top vent at valve A. But this is very wrong. Sandy explained my sinful practice as follows:

1. Connecting the condensate drum via the balance line to the top of the channel head will cause the condensate drum pressure to equalize with the steam supply pressure.
2. The condensate draining from the channel head is equal to the steam supply pressure minus the $\Delta P$ through the tubes. In Figure 7-2 that is

$$100 \text{ psig} - 2 \text{ psi} = 98 \text{ psig}$$

3. The level of water in the channel head has to be 4.6 ft higher than the level in the condensate drum, to overcome the 2+ psi differential head pressure:

$$(2 \text{ psi}) (2.31 \text{ ft/psi}) = 4.6 \text{ ft}$$

4. As the channel head is only 4 ft in diameter, this is not possible without filling the entire tube bundle with water.

"Okay, Sandy, I've made a mistake," I said. "But where's the sin?"

"Norman," she answered, "The sin is that to overcome condensate backup and restore the reboiler duty, my operators must drain the clean steam condensate into the sewer through valve D. Don't you believe that it is sinful to waste fresh water–and boiler feedwater–treating chemicals?"

"So what's to be done?" I asked.

"It's rather simple; connect the balance line at valve B, that is, just below the channel head horizontal pass partition baffle. Then the pressure in the condensate drum will be 98 psig, not 100 psig. Thus, the condensate backup will be eliminated! Got it, Norman?"

"No, I don't understand, Sandy. How can I then vent the $CO_2$ that accumulates in the channel head? Steam always contains a few ppm of $CO_2$. It accumulates at the top of the channel head. If I don't vent it off through valves A and C (Figure 7-2), the $CO_2$ will be trapped. The $CO_2$ must eventually dissolve in the water to form carbonic acid ($H_2CO_3$). This will corrode the exchanger tubes. No, Sandy, I need to leave valve A open and vent from valve C.

"No, Norman! You men all think you are so smart! No! The $CO_2$ must accumulate below the pass partition baffle, after the steam condenses. Then vent it off with valves B and C open. Valve A is closed. Corrosion control on steam systems requires $CO_2$ venting from a high point, after the condensate is collected but not before. Let this be a lesson to you."

"Sandy, what's the lesson?" I asked.

"Learn to respect female engineers. That's the lesson!" Sandy shouted above the roar of the K-802 air blower.

So, Mrs. Hightower, in memory of Sandy Lani, I will apologize again for my idle threats, even though your steam condensate recovery is only 40%, and in other plants it is over 70%. As for Sandy, she is now, I guess, consulting on heavenly hydraulic problems. But, Mrs. Hightower, you didn't look a day over 40 when I saw you at the anniversary dinner. Kindly check the status of the ice cap in Greenland when you reach my age.

## WATER HAMMER

If hot, flashing steam condensate is mixed with subcooled steam condensate in a pipe, water hammer may result. The flashed steam is rapidly condensed as it mixes with the cold water in a localized area of the condensate collection piping. This localized steam condensation then creates an area of localized low pressure. The water in the more remote sections of the piping then rushes into these areas of low pressure. If the accelerated water contacts an elbow or fitting, a crashing or banging sound results. This is called water hammer. Damage to the piping system may well result.

Operators do not, and should not, tolerate severe water hammer. They stop the water hammer either by diverting the hot flashing condensate to the sewer, or by spilling the cold steam condensate onto the deck. Either way, the condensate is lost.

To correct this problem, high-pressure hot steam condensate should be flashed at an intermediate pressure, to remove and recover the steam before it combines with the colder water in the condensate collection piping.

## MEASURING STEAM CONDENSATE RECOVERY

Many of my clients are under the impression that the recovery of condensate is 70 to 80% when it's really 30 to 40%. Determining the percent recovery of condensate is rather simple if two values are known:

- Steam production $= A$
- Treated water makeup to deaerator $= B$

Then I calculate the percent of condensate recovery as

$$\frac{A - B}{A}$$

A reasonably good rate of condensate recovery for a refinery is 70%. However, on several occasions I have calculated for several major refineries that condensate recovery was only 30%.

## COOLING TOWER CYCLES OF CONCENTRATION

As a process operating superintendent in Texas City in the 1970s, I had definite operating targets, one of which was "cycles of concentration" for my alkylation unit cooling water tower. Water is cooled by evaporation. The dissolved solids in the circulating water were supposed to be five times higher than the dissolved solids in the cooling tower makeup water. This means that 80% of the makeup water should evaporate. The other 20% should be drained from the cooling tower basin to purge out the accumulated dissolved solids. This is the cooling tower blowdown. However, I found that even though the blowdown valve was closed, my cycles of concentration were only two, not five. What was my problem?

- The operators were leaving the water running through all the sample coolers at all times.
- Cooling water was sprayed on several product cooler heat-exchanger shells, even though the product cooling was barely improved.
- When cooling water exchangers were back-flushed, this activity went on for an hour, even though the back-flushed water was quite clean after just 5 minutes.
- The cooling water from one critical exchanger did not have sufficient pressure to flow back into the cooling water return header, so the water was drained to the plant sewer.

This last point was the major problem causing low cycles of concentration. I stopped this waste of the cooling tower water by modifying the tube (i.e., water) side of the exchanger from four tube passes to two tube-side passes. This required that the baffle in the floating head be eliminated. The dual off-center baffles in the channel head were replaced by a single central baffle. Note that the channel head tube sheet also had to be re-machined to accommodate the new center-pass partition baffle.

# Chapter 8

# *Incremental Expansion Design Concept: Reprocessing Waste Lube Oil*

I have written some bad things about my clients, the major oil refineries in the United States. For instance, did you know that Essen Oil piped thousands of barrels of visbreaker vacuum tower residue onto the beaches of Carib Island? It was a long time ago. I saw the results of their dumping this heavy hydrocarbon; tons of it was still scattered on the beach in the 1990s.

Ataco Oil disposed of their waste $AlCl_3$ hydrocarbon complex from their Arkansas viscous polypropylene plant by running the waste down a 4-in. pipe into a small valley, which in Arkansas they call a "hollow." I helped design the plant, so I know.

The Reef refinery in Carib Island stripped ammonia sulfide out of sour water so that the stripped sour water could be reused as desalter wash water (see Chapter 7). Great! But the stripper overhead reflux drum ran at $190°F$ and floated on the flare pressure. At $190°F$, the vapor pressure of water caused the $NH_3$ rich off-gas to be 50% steam.

The mixture of $NH_3$, $H_2S$, and $H_2O$ vapor ran down to the flare knockout drum through 500 ft of 8-in. bare pipe (Figure 8-1). By the time the vapors had reached the flare knockout drum, ambient heat losses had condensed most of the water vapor. The water condensate then reabsorbed most of the $NH_3$. The renewed sour water was then drained from the flare knockout drum through a 2-in. rubber hose that drained into the sea 50 ft away. The local outside operator told me that the hose was in its current location when he was hired 12 years ago.

*Process Engineering for a Small Planet: How to Reuse, Re-Purpose, and Retrofit Existing Process Equipment.* By Norman P. Lieberman

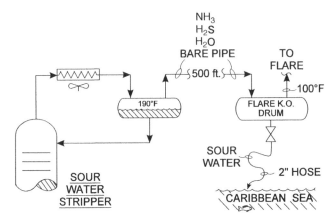

**Figure 8-1**  Ineffective operation of a sour water stripper.

## RESURRECTION

And then we have the tale of the Texaco Marine Division. There must be a special place in heaven for these sainted folks. Verily, they shall be rewarded in heaven for their earthly deeds. For they hath resurrected valuable hydrocarbon base stocks from used lubricating products and discarded motor oils using vacuum distillation.

Their plant is located on the west bank of the Mississippi River, just upriver from New Orleans. The modern Texaco waste oil recovery plant stands out in stark contrast to the decaying and abandoned structures lining this section of the river. The Texaco Marine Division collects contaminated lubrication products along the river for transportation to their plant. It is only a 1500-bsd facility, hardly worth the effort, considering Texaco has to pay for delivery of all their contaminated feedstock. If this were really meant to be a for-profit operation, Texaco would have built their plant 40 miles upriver at their giant Convent, Louisiana refinery.

No, it's not earthly profit that the Texaco Marine Division seeks. It is a heavenly reward that they shall surely receive for reducing the dumping of used lube oils into the Mississippi River.

## A VISIT FROM MR. ROBERT WHITE

I helped revamp a vacuum tower across the river for Mr. White using my "incremental expansion design" concept. One morning I came home and found Robert White sitting at my kitchen table drinking tea and talking to Liz about the benefits of a vegetarian diet.

Mr. White rushed up to shake my hand with a firm and friendly grip. "Norman, great to meet you. I live just down the next street. 'Why not just pop by?' I said to myself."

"Really," I said, "I was just running on the levee by the river. I'm kind of sweaty."

"Oh! Don't worry about a little sweat. It's fine. You can see our plant across the river, you know. I'm the plant manager of the Texaco marine waste oil recovery plant. It's a new plant. We just started up last year. Hope you don't mind my visiting you and Liz so unexpectedly?"

"Yeah, I've seen your plant. How's it running?" I asked.

"Norman, I've got copies of all your books. I keep your *Troubleshooting Process Operations* on my desk. I just thought 'Here is the man who can help us. And he's a neighbor, too! What luck.' So I just popped in for a cup of tea and here I am," Robert laughed.

"Help with what," I asked.

"Norman, it is so kind of you to ask. And Liz, thanks for the tea. Basically, Norman, we need help with everything. We are only running at half rate, half the time. The longest run we have had is 37 days, and none of our products are on spec. Our feed tanks are full. Worst of all, Texaco Marine headquarters in New York will only cover my operating losses for another six months." Mr. White paused, "Then what will happen to all that waste oil? Can you help me?"

"If you have the money, I have the time," I responded.

"Great, Norm. I'll pick you up at 8. Let's have breakfast at the chic café across from Jackson Square. Their croissants are really good."

## THE TEXACO MARINE DIVISION VACUUM TOWER DESIGN

"Mr. White," I asked the next morning, "Who designed this plant for you? Why not ask them for help?"

"Oh, Norman, please call me Robert. It was designed by the Texaco refinery design division in Houston. We in Texaco Marine just thought that vacuum tower design was a refinery thing. And now look! It's such a failure, and they don't seem to care. They said the plant is designed correctly, but we don't know how to run it. It's frustrating! You just can't talk to those people. I've phoned and phoned, but they don't care."

"Robert, let's sit down in the control room, talk to the unit operators, and make a list of the problems as they see them."

"Wouldn't it be best to review the Texaco computer model of the plant first? We paid $250,000 for the Hysim model. I think it must be a valuable design tool," said Mr. White.

"Robert, it is a little late for fairy tales. Let's go talk to the operators."

I always start my incremental expansion design program by first establishing a list of known operating problems, as perceived by the shift operators. In this case, my list included (see Figure 8-2):

1. Both the LVGO (light vacuum gas oil) and HVGO (heavy vacuum gas oil) products were black.
2. The wash oil packing coked-up solid every few months.
3. The vacuum was bad at the top of the tower and in the flash zone.

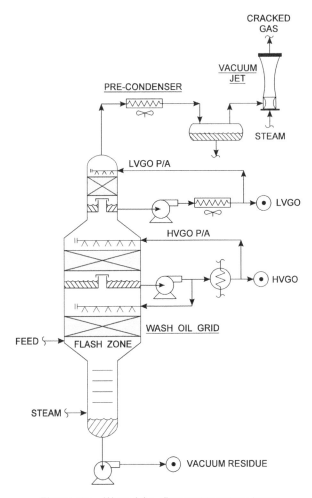

**Figure 8-2** *Waste lube oil recovery vacuum tower.*

4. The bottom's pump cavitated at design rates.

5. The feed preheat exchangers fouled very quickly and very badly.

6. The vacuum tower bottoms product did not meet flash specifications.

7. Vacuum heater fuel consumption was excessive.

8. The LVGO pump-around rate was limited to 70% of the required rate.

I did not intend to expand the capacity or efficiency of the unit as a whole. Rather, my method would be to correct the known problems that seemed to the plant management and the unit operators to be the most critical. Some problems could be corrected on-stream and some problems would require a plant shutdown to correct. Some problems could be corrected by modifying existing equipment. Others would require purchase of new process equipment.

## WASH OIL GRID COKING

I began with the problem of the wash grid coking. When this packed section partly plugs with coke (see Figure 8-2), localized velocities become high. This leads to entrainment of the black feed up into the HVGO product draw-off tray. The HVGO product then turns black due to contamination with feed. The circulating HVGO pump-around is distributed through a spray header. Some liquid HVGO is always turned into a mist in the spray header spray nozzles. The mist is entrained up into the LVGO draw-off tray, which then turns the LVGO draw-off black.

The HVGO and LVGO products were intended to be sold to the Texaco refinery as FCU feed, for conversion into gasoline and diesel products. But the refinery would not accept black FCU feed, as it contained cracking catalyst poisons. Thus, the HVGO and LVGO products were blended back into feed tankage, which defeated the entire purpose of the operation. The problem was the history of the feed. It contained the degraded lube oil additive package, thermally degraded hydrocarbons, and oxygenated components. That is, the slightly acidic sludge we have all seen when we drain our car crankcase and change the oil filter.

Apparently, the sludge was entrained from the vacuum tower flash zone onto the wash oil packing. Being more thermally unstable than virgin hydrocarbon components, it coked on the wash oil grid. Once fouling begins on any packed bed, a positive feedback loop is created as follows:

- A localized high vapor $\Delta P$ is created.
- The high $\Delta P$ increases the residence time of liquid in the packing.
- Thermal degradation (coking) is a function of time and temperature. The greater liquid holdup in the packing thus promotes more fouling due to coking.
- The problem builds upon itself.

My solution to this problem was rather simple. First, let me define the *entrainment velocity*, or *C* factor:

$$C = V \left( \frac{D_V}{D_L} \right)^{1/2}$$

where
$V$ = superfacial vertical vapor velocity, ft/sec
$D_V$ = vapor density, lb/ft$^3$
$D_L$ = liquid density, lb/ft$^3$

Assuming that the feed is reasonably distributed in the flash zone, my experience teaches the following:

- $C = 0.16$ ft/sec or less: gravity settling is sufficient to suppress entrainment.
- $C = 0.25$ to 0.35 ft/sec: a grid or demister is required to promote deentrainment.
- $C = 0.45$ ft/sec: entrainment is uncontrollable regardless of internals.

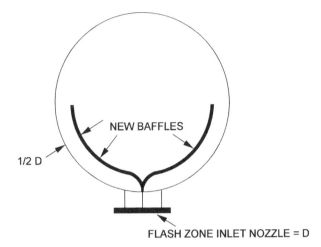

FLASH ZONE INLET NOZZLE = D

**Figure 8-3** *Converting a radial inlet nozzle into dual tangential entries for better feed distribution.*

The $C$ factor calculated in the vacuum tower's flash zone was slightly above 0.16 ft/sec. Thus I reasoned, the wash oil grid was doing more harm than good. I would remove the wash oil grid completely, but continue to use the wash oil spray nozzles to help knock back any residual components.

I have found that using a well-designed spray header will, at the very best, suppress only half of the entrained residual components. I retrofitted this spray header according to the following process:

- Select a spray angle so that not more than 10 to 15% of the wash oil spray hits the vessel wall above the feed inlet.
- Use Bete full-cone spray nozzles [1].
- Design for 200% spray coverage at the elevation just above the feed inlet distributor.
- Use hot HVGO rather than the cold pump-around return.

To ensure a reasonable vapor distribution in the flash zone, I added the dual tangential baffles shown in Figure 8-3. If feed inlet velocities are high (above 100 ft/sec), a tangential entry acts somewhat like a centrifuge, forcing entrained liquids toward and down the vessel wall.

## STRIPPING TRAY EFFICIENCY

During the same turnaround in which the black HVGO and LVGO problem was corrected, I also corrected the low flash problem of the vacuum towers bottom product. Flash is a measure of how easily a hydrocarbon will ignite when exposed to a flame. The lower the temperature at which a hydrocarbon can be ignited, the lower the flash point temperature. A low flash-point temperature is a result of excessive light hydrocarbon components. Such lighter components are removed from residual fractions by steam stripping. The problem with the steam stripper was excessive hole area in the tray decks. For trays to work correctly, the pressure drop of the vapor flowing through the holes must be approximately (+ or –50%) equal to the weir height (typically, 2 to 3 in.). To calculate the $\Delta P$ of the vapor flowing through the sieve tray holes used in this tower, I used the following formula:

$$\Delta P = 0.3 \, \frac{D_V}{D_L}(Vg)^2$$

where
    $\Delta P$ = pressure drop of vapor flowing through the sieve hole, in. of liquid
    $V_g$ = vapor hole velocity, ft/sec

The existing trays had a hole $\Delta P$ value of only 0.3 in. of liquid, and a weir height of 3 in. Hence, liquid dumped down the tray decks, rather than overflowing the weirs into the downcomers. As the downcomers were empty and unsealed, the stripping steam followed the path of least resistance and flowed upward through the downcomers. All this destroyed the vapor–liquid contacting and tray stripping efficiency.

To correct this problem, I had two-thirds of the sieve holes on each tray deck closed with $\frac{1}{2}$-in. bolts. Further, I had a steam trap installed on the stripping steam supply line to remove excess moisture. Wet steam robs the oil phase of heat and thus retards stripping efficiency. Both changes combined resulted in an on-spec, vacuum tower bottoms product flash point temperature. Thus, this product could now be sold for industrial fuel oil cutter stock.

## PRECONDENSER FOULING

If the stripping tray hole area was excessive, why not just use more steam? Well, because the vacuum jet air-cooled precondenser (Figure 8-2) was fouled and had an excessive pressure drop. It was badly fouled and could not be cleaned. For some unfathomable reason, the air cooler tubes had been purchased with spiral spring inserts to induce turbulent flow. This would reduce the heat transfer film resistance and increase the cooler's heat transfer coefficient. At least that's the theory. I guess it worked for a while; but once fouled, the long springs could not be extracted from the tubes.

Admitting defeat on this front, I spec'ed-out a new tube bundle, but without the springs. Ladies and gentlemen, vendors have been marketing various sorts of tube inserts since I was an entry-level engineer in the 1960s. If they were really much good, don't you think they would be in widespread use by now? The new tube bundle was a six-month delivery item, but it did reduce flash zone pressure and increase the HVGO recovery when it was eventually installed.

## PUMPING PROBLEMS

Actually, I had an alternative method of reducing the precondenser vapor load. This was to reduce the tower top temperature by circulating more LVGO. The LVGO pump-around pump was limited by cavitation. That is, above 600 bsd of flow, the pump suction pressure was not sufficient to provide adequate NPSH (net positive suction head) for the little centrifugal pump. The problem was that the height of the chimney on the LVGO draw-off pan was too short. The minimum height needed for this chimney, as measured from the middle of the draw-off nozzle, is

$$H = 0.34(V_L)^2$$

where

$H$ = height of liquid, in.
$V_L$ = velocity of liquid through the draw-off nozzle, ft/sec

Note that the coefficient 0.34 excludes any frictional loss of piping and fittings located at the same elevation as the nozzle. This means that the draw-off pump suction line should elbow down from the nozzle ASAP.

I threw away the existing chimney hat and increased the chimney height from 1 to 2 ft. The maximum LVGO P/A flow increased by about 40%.

The vacuum tower bottoms pump limitation proved to be insurmountable. The evil engineer who set the elevation of the vacuum column will meet his (or her) just retribution someday soon. The error was that the vacuum tower was too close to grade. To achieve the design unit capacity, I had to run both the primary and spare bottoms pump. This undesirable operating practice reduced the required NPSH by over one-half. Most of the NPSH required of a pump is associated with accelerating the flow into the eye of the impeller. Since the potential energy (suction head) needed to accelerate a flow varies with the flow squared, running two pumps instead of one is an easy way out of a running NPSH requirement limitation. Within a year, the Texaco Marine Division installed a new bottoms pump that required very little NPSH, as it had a huge impeller eye.

## EXCHANGER FOULING AND HEATER OVERFIRING

To suppress the preheat exchanger fouling, I called a fellow process engineer, Jim, who runs a waste lube oil recovery unit in Chicago. This is also one of my best techniques to overcome process limitations cheaply. I do not steal information: I borrow information and return it promptly after use. From Jim, I learned that the fouling tendency of acidic lube oil sludge may be reduced by neutralizing the organic acids formed by oxidation of the used lube oil. At the time, the Texaco refinery in Convent, Louisiana, had a large surplus of waste caustic produced from sweetening sour (i.e., high-mercaptain) gasoline that the Texaco Marine Division could use to neutralize the waste lube oil feed.

The resulting reduction in exchanger fouling improved the feed preheat temperature and thus reduced the firing rate of the preheat furnace. A further reduction in firing rate was achieved by finding and fixing the tramp air leaks in the preheat furnace convective section, in accordance with the following formula (see Chapter 3):

$$\Delta F = \frac{(\Delta O_2)\,(\Delta T)}{500}$$

where

$\Delta O_2$ = oxygen difference of the flue gas measured above and below the convective section, %.

$\Delta T$ = temperature difference between the stack and the ambient air, °F

$\Delta F$ = percent of heater fuel wasted due to cold tramp air leaks into the convective section tube banks, %

The improved heat exchanger and fired heater combustion efficiency reduced the fuel gas consumption required by almost 10%.

The West Bank River Road runs along the Mississippi River to New Orleans, past once-proud but now forlorn industries. It's sad. But as I drive past the Texaco Marine vacuum tower, I feel uplifted and fulfilled. A huge orange sign above a gray tank beckons:

"DEPOSIT WASTE LUBRICATION OIL HERE"

It's a nice feeling to know that I played a part in helping those wonderful operators, engineers, and managers from the Texaco Marine Division in White Plains, New York, in saving our little green planet from used lube oil contamination.

(*Note:* Unlike my other references to refineries, which are fictitious, the reference to the Texas Marine Division is accurate, as is this story.)

## TRANSFER LINE SONIC VELOCITY: MYTH OR REALITY?

One of the things I decided not to do at the Texaco waste oil reprocessing facility was to increase the diameter of the fired heater transfer line. This line connects the heater to the vacuum tower.

I had calculated that the velocity in this transfer line was more or less at sonic velocity, or the speed of sound. The speed of sound in air is about 1000 ft/sec. The speed of sound varies with molecular weight and the temperature of the vapor phase. Very approximately, sonic velocity is proportional to the square root of the vapor density.

According to the theory advanced by both major providers of vacuum tower internals, transfer line velocities, or even velocities in the vacuum heater tubes, in the range of sonic velocity (roughly 500 ft/sec under vacuum conditions) promote severe entrainment of residual components. According to their presentations, when a vapor stream reaches sonic velocity, entrained droplets of residual components will shatter into a finely divided mist. The tiny droplets of mist cannot be settled out by gravity in the vacuum tower flash zone because they are too small. This problem explains, according to vendor claims, why severe entrainment persists in some vacuum towers, even though the flash zone entrainment velocity (i.e., the $C$ factor) is quite moderate, at 0.30 to 0.35 ft/sec.

But I have had my doubts. I well recall standing on the truly gigantic new vacuum heater transfer line at the Baxter refinery in Los Angeles with Tom Caine. Tom was still angry.

"Norm," he fumed, "those idiots convinced me to change our vacuum tower feed line from 48 in. to 78 in. Also, I increased the size of the last two tubes in the vacuum heater passes from 12 in. to 18 in. Why? Well, they claimed that I needed to reduce the formation of finely divided droplets of vacuum resid. This was supposed to clean up my vacuum gas oil and suppress entrainment of asphaultines."

"So, I guess it didn't help that much, Tom."

"It didn't do a thing," Tom screamed above the roar of the heater's burners. "We wasted $500,000 on this nonsense!"

So when faced with the same claims, by the same vendors, for my project at the Texaco waste oil recovery plant, I decided to analyze some data from a similar vacuum tower in Kansas. These data showed clearly that good-quality vacuum gas oil was produced, even at transfer line velocities that were in the range of sonic velocities, as long as the vapor horn, the wash oil packing, and the other vacuum tower flash zone internals were well designed. So I phoned my contacts with the two major tower internal vendors, and they told me that in reality there were no firm data to support their claims that sonic velocities caused unstoppable entrainment of black components in the flash zone of vacuum towers.

But then, I wondered, why were these same vendors marketing this technology all over our poor little planet if they did not have any real data to support their claims? Could it be they were trying to market engineering studies based on this somewhat exaggerated idea of shattering liquid droplets at a calculated sonic velocity?

Just think about all the steel and energy that's wasted replacing vacuum heater tubes and vacuum feed transfer lines based on the claims of the exaggerated effects of sonic velocity in vacuum heater transfer lines and furnace tubes. Isn't engineering supposed to be an ethical science?

## SUMMARY

Having designed process equipment for many years and revamped hundreds of process units, I have found it best to fix those problems that are obviously limiting production. Trying to anticipate all possible limitations and revamping all parts of the plant at the same time typically results in wasting capital investment. Other than tower internals, most process limitations can be fixed on-stream. To be successful in this incremental expansion design approach to revamping process units, the designer has to have an intimate knowledge of the unit and the process. This knowledge is best gained through discussions with shift operators and by personally collecting on-site field data. This is the "minimalist" approach to process engineering, which is what this book is all about.

As you read my other books (see the preface for a list), you can see that I have always tried to avoid adding more process equipment to achieve my clients' objectives.

In a political context, the two methods to reform a government are evolution and revolution. Democracy is a system that promotes evolutionary political change. Communism is a system that promotes revolutionary change. My incremental design expansion method is akin to the evolutionary political progress we are able to achieve in democratic societies when we exercise patience and the knowledge derived from experience.

## REFERENCE

1. *Bete Spray Nozzle Catalog* (Full Cone Nozzle Section, p. 34).

Chapter *9*

# Improving Fractionation Efficiency in Complex Fractionators

Why would Zeus select me to be his messenger? After all, why not Hermes? Certainly, I am an experienced process design engineer. Perhaps my position is not that unique. Zeus has probably selected thousands of people, each working toward a common goal: to save the biosphere from humankind's excessive and careless consumption.

Perhaps my efforts to increase refinery efficiency is really divinely inspired. For example, in 1968, I was demoted from senior design engineer to the position of subassistant tech service engineer on the No. 12 pipe still. This was a 220,000-bsd crude unit in the American Oil refinery in Whiting, Indiana. I secured this demotion in a quite clever way. Having been promoted to the long-range strategic executive committee, I threatened to resign unless I was transferred to a tech service position. As a punishment for my lack of appreciation for the promotion, I got my subassistant job at Whiting.

In my new position, I reported directly to Jerry, a failed goat herder from El Dorado, Arkansas. He had become a chemical engineer, but his heart was with his goats in the green valleys of Arkansas.

"Norm, you work on the design stuff. I'll work with Pete on the day-to-day stuff," Jerry suggested.

"What design stuff?" I asked.

"You know, stuff to save energy. Stuff to improve fractionation. Ever since that Guy Walker revamped the unit, fractionation is real bad, especially between light straight run (LSR) and virgin heavy naphtha reformer feed (HVN)" (see Chapter 4).

*Process Engineering for a Small Planet: How to Reuse, Re-Purpose, and Retrofit Existing Process Equipment,* By Norman P. Lieberman
Copyright © 2010 John Wiley & Sons, Inc.

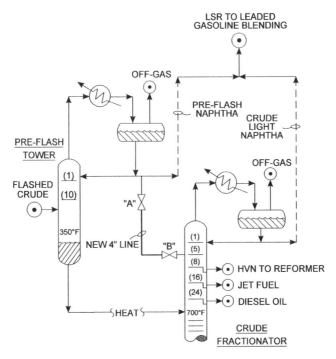

**Figure 9-1** *Use of pre-flash naphtha as an intermediate reflux in crude tower to improve LSR-HVN fractionation.*

In a sense, Jerry was right; fractionation at the No. 12 pipe still was bad. For example, the LSR was contaminated with 25% HVN (see Figure 9-1), and HVN components had a very low lead susceptibility.

## LEADED GASOLINE

The awful story of the use of tetraethyllead (TEL) marks a black time in the history of the refinery industry in the United States. The technology to produce gasoline of the required octane was available long before the government-mandated phase-out of leaded gasoline in 1982. The only components of gasoline to which it really made economic sense to add lead were:

• Normal butane
• Normal pentane
• Normal hexane

Isopentane and isohexane already had reasonably good octanes. Ring-type $C_6$ components are called *naphthenes*. These rings are also called *benzene precursors*.

The naphthenes, $C_7$'s, $C_8$'s, $C_9$'s, and lighter $C_{10}$'s had poor "lead susceptibility." One therefore had to add a lot of TEL to make a significant octane improvement.

Normal butane has a high vapor pressure (RVP) and should not have been included in the huge concentrations used to blend into gasoline (sometimes up to 10 mol%), because much of the butane evaporated during and after flowing into the customer's gas tank. Normal pentane and normal hexane are both present in fairly high concentrations in crude oil, and both really increase greatly in octane when 3 grams of TEL per gallon (1000 ppm) are added.

It is not as if the refining industry did not understand the toxic nature of TEL. Handling of TEL at the refineries was extremely cautious; operators who blended in the lead wore special protective clothing and gloves. Certainly, the technology to enhance the octane of normal pentane and normal hexane by isomerization to isopentane and isohexane was well established. In 1965, the technology to convert normal butane to isobutane, using the same process configuration as that of a modern pentane–hexane isomerization unit, was available. However poisonous, toxic lead was cheaper than isomerization.

Leaving hydrocarbon components in the LSR heavier-than-normal hexane partly defeated the purpose of adding TEL to gasoline. Such heavier components, especially the ring-type naphthenes, were supposed to flow to the "naphtha reformer" for conversion into benzene. That is, we tried to maximize the benzene content (a carcinogen) in gasoline.

Until the government put a stop to it, all the major refiners happily blended gasoline with carcinogenic benzene, volatile butane, and toxic lead. And my retrofit design at the No. 12 pipe still recovered low-lead-susceptibility benzene precursors from LSR for naphtha reformer feed (HVN).

## FRACTIONATION PROBLEMS WITH A CRUDE UNIT PRE-FLASH TOWER

My co-worker, Guy Walker (see Chapter 4), had expanded the capacity of the No. 12 pipe still by adding the pre-flash tower shown in Figure 9-1. Pre-flash towers recover 10 to 15% of the crude as the pre-flash naphtha. This is good for two reasons:

1. The energy needed to heat the crude charge is reduced. About 85% of the crude must still be heated to 700°F. But as shown in Figure 9-1, the pre-flash naphtha has only to be heated to 350°F.
2. The vapor velocity in the crude fractionation tower is reduced. Although only 10 to 15 liquid volume percent is flashed off in the pre-flash tower, this may represent 20% of the moles of vapor in the crude tower. This is a good route to "de-bottleneck" a crude fractionation tower.

The factor that reduced both the energy and capacity benefits of the pre-flash tower was the reflux rate required to control the LSR endpoint. The LSR could not be separated from the HVN without trays and reflux. But using a lot of top reflux reduced the amount of LSR recovered in the pre-flash tower. Using too little reflux increased HVN losses to LSR. This wasted lead (TEL) and reduced the benzene content of blended gasoline. Using too much reflux defeated the purpose of the pre-flash tower.

Guy Walker had used an optimum reflux ratio in his design. But Pete, the operating manager for the No. 12 pipe still, found that he could run an extra 10,000 bsd of crude, by running at a very low (1 : 4) reflux-to-LSR product rate. But this low reflux ratio (1:4), compared to the design (1 : 1) reflux ratio, resulted in about 25% HVN left in the LSR product.

## INTERMEDIATE REFLUX CONCEPT

I rather liked the idea of running at the low 1 : 4 reflux rate, as it saved energy in the crude preheat furnace. Pete had proposed that American Oil construct a new naphtha splitter to rerun the pre-flash tower naphtha and recapture the 25% HVN content for reformer feed. However, even four decades ago I rebelled against the very thought of a new distillation tower. It just seemed to be the easy way out of a problem: like Adam eating the apple because he didn't want to get into a big argument with Eve. Fortunately, Jerry provided the solution to this quandary.

Jerry had caught pint samples of both crude light naphtha and pre-flash naphtha. Perhaps thinking they were fermenting goat milk, he had left both bottles open on his windowsill. The next morning the crude light naphtha bottle was dry, but the pre-flash naphtha was still about 20% full. I might have accelerated the process by dropping a pebble in each bottle. The pebble serves as a "boiling stone."

The ASTM D-86 distillation of the crude light naphtha (Figure 9-1) confirmed that the crude light naphtha was almost totally free of HVN or benzene precursors. This gave me an idea. Suppose that I was to run a 4-in. line from point A to point B, shown on Figure 9-1. Point A was the discharge of the spare pre-flash tower reflux pump. Point B was an idle naphtha draw-off nozzle above tray 5. This would allow the HVN content of the pre-flash naphtha to be recovered in the crude fractionator HVN product stream.

Of course, the top reflux rate on the crude fractionator would also be reduced in proportion to the pre-flash naphtha flow rate. That is, the pre-flash naphtha would serve as an intermediate cold reflux. The result would be a decrease in fractionation efficiency on the top four trays in the crude fractionator. Certainly, then, the amount of benzene precursors (HVN) in the crude light naphtha stream would be higher than the current operations. But the HVN lost to the pre-flash naphtha would be zero, because no separate pre-flash naphtha product would be exported to the leaded gasoline blending pool.

In the late 1960s, we did have refinery distillation computer models. But there was only one computer available for the Whiting refinery, Central Engineering, and R&D, and a subassistant engineer to a goat herder was not permitted access to this lofty machine. And then it was only 200 ft of 4-in. carbon steel pipe that was needed.

So I guessed that the net result would be to eliminate 70% of the HVN content of the LSR product.

Then I told Jerry and Pete that my creative spirit had inspired me and that it would be insulting to ask to see my calculations.

"Couldn't you get any computer time?" asked Pete. "Too expensive?"

"Sure I could, but I didn't need it. The intermediate reflux concept is divinely inspired. Oh ye of little faith," I concluded.

Only half-convinced, Jerry asked, "Norm, this ain't no Yankee spirit that's up and inspired you, is it? You're talking about a genuine rebel-type spirit!"

"Absolutely, Jerry, that very one."

## PROJECT IMPLEMENTATION

I really did guess at the incentive for this project. The computer time for me to run my model would have cost $1000. Pete added a flow meter and a control valve to my 4-in. line shown in Figure 9-1. Jerry returned to southern Arkansas to raise chickens for Tyson Frozen Foods. Pete joined the strategic energy task force. The American Oil Company has disappeared into the mists of time. My intermediate naphtha reflux project for the No. 12 pipe still was forlorn and long forgotten for four years. Suddenly, it, too, was resurrected. The strategic task force committee later estimated that the intermediate reflux project paid for itself in six days. But no one, until just now, remembered that it was I who had first inspired the project, four long years before it was finally executed.

## DIESEL RECOVERY FROM A VACUUM TOWER FEED

One day shortly after our pre-flash tower intermediate reflux project meeting, Pete rushed into the cubicle I shared with Jerry Edwards.

"Lieberman!" Pete shouted. "Design me a diesel oil recovery tower! Quick! Pronto! Today!"

"A what?" I asked.

"Listen up," Pete answered. "We have 14% 650°F minus stuff in the crude fractionator bottoms. That 650°F minus stuff is diesel. The new diesel oil recovery tower will refractionate the crude tower bottoms before we lose the diesel to vacuum tower feed. Once the diesel gets into the vacuum tower, it winds up in LVGO, which goes to FCU feed."

What Pete Markowitz was trying to say was that an ASTM D-86 distillation analysis of a sample of the vacuum tower feed, showed 14% distilled off at 650°F. The 650°F point somewhat understates the diesel oil content of the sample. I knew what had happened. Pete's girlfriend, Lydia Rabowski, the operations manager of the FCU complex, had complained to him that she was getting too much light material in her feed. Lydia was right, but Pete was overreacting.

"So how's Lydia?" I inquired in a friendly tone, "you two planning to get married?"

"Mind your own business," Pete was angry because both of us knew that he was more or less already married. "I need a process design for the new tower yesterday. Cost! Not to exceed $5,000,000. Feed rate! 95,000 bsd. Spec's! Bottom's not to exceed 6%, 650°F and lighter. Diesel endpoint, 680°F. Diesel flash, not less than 140°F. Got it Lieberman? Any questions?"

"No. A okay," I acknowledged.

"Jerry," Pete continued slightly calmer, "go find a nice plot plan location for the new tower. Figure 10 ft I.D. and 70 ft T-T (tangent to tangent). Let's get started boys. This ain't no prayer meeting."

Pete stormed out. The "prayer meeting" reference had offended Jerry. The "diesel oil recovery tower" reference had offended me. The idea of rerunning crude fractionator bottoms in a new tower was degrading, a humiliating and insulting suggestion. My integrity as a chemical engineer and process designer had been attacked. I knew, like everyone, that I was created for some particular purpose—and that purpose was not to rerun the crude fractionator bottoms in a new diesel oil fractionator.

I ran after Pete, "Do you mind if I just check the $\Delta P$ across the crude tower bottom six stripping trays? Maybe they're screwed up. You know, Pete, a lot of the diesel is stripped out with the steam on these trays." (See Figure 9-2.)

"Listen, Norm, I inspected those trays in the May turnaround myself. They were in perfect condition. Didn't you design those trays yourself when you worked in the process design division?" and Pete added, unnecessarily, "before your most recent demotion."

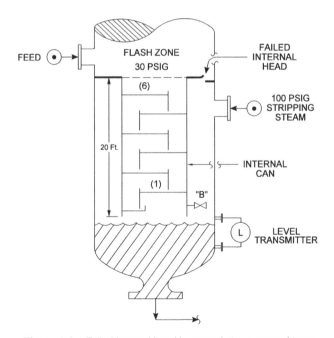

**Figure 9-2** *Failed internal head bypassed steam around trays.*

"Pete, maybe we blew out the tray decks since May" I suggested defensively.

"Maybe your design was no good," Pete responded. "Look, Lieberman. We have had high diesel losses into tower bottoms for 20 years. Same with your new trays. Same with the old trays. Same before and after the May turnaround. Okay. Go ahead and waste more time. Measure $\Delta P$ across your stripping trays. But then, get on with my design for the diesel recovery tower. Lydia's been bugging me every night about diesel in her cat cracker feed. Also, she's late this month," Pete added in a softer and vaguer tone.

## BYPASSING STRIPPING TRAYS

About one-third of the crude towers I have worked with have the feature shown in Figure 9-2. The bottom six trays are contained inside a "can" that is internal to the crude tower itself. The tower bottoms liquid level is maintained below the bottom edge of the can. But should the bottoms liquid level rise up above the can, the stripping steam will be trapped in the annular space between the can and the crude tower vessel wall. The pressure in this annular space will rise until the entire 20-ft-high can is full of liquid. That suggests that a differential pressure equivalent to 20 ft of liquid (or 7 psi at a specific gravity of 0.80) would develop across the internal head. But the internal head was constructed to resist an uplift force of only 1 psi, not 7 psi. After all, the internal head was constructed of a flat plate of 12-gauge steel clipped to the tray ring on the vessel wall. Later inspection showed that this plate pulled and bulged away from the tray ring, as shown on my sketch.

But first, I made a field measurement of the differential pressure between the flash zone and the top of the level transmitter. The $\Delta P$ calculated between these two points was 1.2 psi, the $\Delta P$ observed was zero. Even increasing the stripping steam rate from 14,000 lb/hr to 21,000 lb/hr failed to increase the differential pressure observed between the flash zone and the level transmitter.

Using the vessel sketch as a guide, I raised the bottom liquid level to the bottom edge of the can. This should have caused my lower pressure gauge to jump 7 psi in just a few minutes. But both my pressure gauges read 30 psig the entire time.

Also, I noted that varying the stripping steam rate had no effect on the diesel oil content of the crude fractionator bottoms product. Combined with the zero $\Delta P$ across the bottom six stripping trays, I felt confident that I had discovered why the stripping tray efficiency was essentially zero. That is, the stripping steam was internally bypassing the trays.

## EFFECT OF USING BOTTOM STRIPPING TRAYS

"Pete," I explained the next day, "Our Sugar Creek refinery runs the same crude as us, at the same flash zone temperature and pressure. But they only have 5%, 650°F, and lighter in their vacuum tower bottoms."

"Get to the point, Lieberman. I'm real busy."

"My point is that Sugar Creek doesn't need a special extra tower to refractionate the crude bottoms to keep diesel out of cat feed. My point is that their bottoms steam stripper is working and ours is not. My point is we have a hole punched in the internal head below the flash zone."

"What? What hole?" Pete stared at his plant phone, barely listening to me.

"The hole that you found in the internal head during the May turnaround," I said. "What's up? Are you okay?"

"Oh, it's nothing. Just something with Lydia. But we fixed that hole in May." Pete seemed transfixed by his phone. "It's just that Lydia's late."

"Yeah, but Pete. We have found that same sort of hole in the internal head for 31 years. It's been reported every shutdown and every turnaround. That's nine turnarounds."

"Nine what, Norm?"

"Pete, are you listening to me? Nine times that hole has been found in the internal head. Any minor high liquid level will seal off the bottom edge of the internal can. Then the steam pressure gets trapped in the annular space between the can and the vessel wall. The steam pressure tears the internal head away from the vessel's supporting tray ring. Then the steam blows right past the trays. The stripping tray efficiency is zero. That's why we have all that diesel oil in the FCU feed. Pete, are you listening, or what?"

The plant phone suddenly rang. Pete grabbed the receiver before the ring faded.

"Okay, Lydia," he whispered. "Yes, I understand. Okay."

Pete's face was pale. He seemed to have trouble breathing. His hand trembled as he replaced the receiver.

"Was that Lydia? Was she asking about the diesel oil content of her cat feed? I bet she will be pleased that we can fix the problem with a quick shutdown."

"Yeah Norm, Lydia's sure to be pleased. She's got a super grasp of refinery overall economics. Did you know she received an A-1 plus overall rating last month in her annual performance review? Lydia's real smart. Did you know that, Norm?"

"Yes, I do. So I'll submit a sketch showing the modification to the stripping section. I'll just connect the steam inlet nozzle to a new valve B (as shown in Figure 9-2), welded to the can below the bottom tray. I'll design the internal steam line for thermal flexibility. Also, I'll put a new steam sparger pipe distributor below tray 1. Then even if you get a high level, the steam pressure won't get trapped between the can and the vessel wall. The steam will just keep flowing up the trays."

"Right, Norm, that sounds great," Pete said, but he still seemed distracted.

"And can I then forget about the diesel oil recovery project?" I asked.

"What? What project? Yeah, forget about it, Norm. I'm sure you know best. That design stuff is your thing," Pete said in a shaking voice. "Listen, Norm, did Jerry leave yet? I need to talk to him. It's personal."

"He leaves early on Thursdays for his prayer group encounter. You know, that little white church on Indianapolis Blvd., down by the lake."

"Yeah, the white church," Pete mumbled as he rushed out of his office.

When we started back up a few months later, the 650°F and lighter ASTM D-86 distillation point of the crude tower bottoms stream was just 5%. The stripping steam was connected to valve B on the internal can (as in Figure 9-2), and all was well. But not with Pete Markowitz. He transferred out of the Whiting refinery before the No. 12 pipe still turnaround. He became a strategic planner in corporate headquarters before he was again transferred to Karg Island in Iran.

## IMPROVING FRACTIONATION USING PICKET WEIRS

Figure 9-3 shows a section of a tower with two pump-around or heat removal sections, plus one fractionation zone.

- Trays 10 to 12 are light distillate pump-around.
- Trays 13 to 16 are fractionating between light and heavy distillate.
- Trays 17 to 19 are heavy distillate pump-around.

The liquid rate on pump-around trays is typically 10 times higher than on fractionation trays. This creates a problem of low weir loading on the fractionation trays. By "weir loading" I mean the flow of liquid over the weir (in hot, U.S. gpm), divided by the length of the weir (in inches). Weir loadings in excess of 12 to 14 gpm/in. cause excessive entrainment. Thus, the weir loadings on trays 13 to 16 will be only about 1 gpm/in. This low weir loading creates poor tray efficiency in several ways.

Low weir loadings result in a low crest height. Crest height is the height of liquid overflowing the top of the weir. For example:

- A weir loading of 1 gpm/in. produces a crest height of 0.4 in. of liquid.
- A weir loading of 12 gpm/in. produces a crest height of 2.1 in. of liquid.

A small out-of-levelness in the weir, when the crest height is small, will cause all of the liquid to spill across the lower portion of the weir. The liquid behind the higher

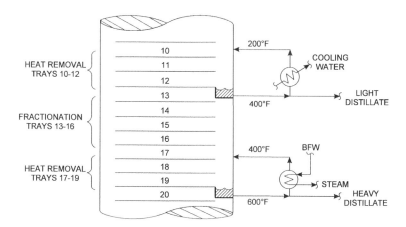

*Figure 9-3*    *Pump-arounds affect internal reflux rate.*

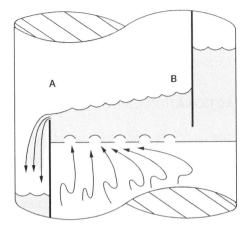

**Figure 9-4**   *Low weir loadings promote vapor channeling.*

portion of the weir will then be stagnant. As vapor flows through stagnant liquid, its composition cannot change. Thus, the section of the tray behind the higher portion of the weir has a tray efficiency approaching zero.

Next, consider Figure 9-4. Note that the liquid level on the tray near the outlet weir is lower than the liquid level at the opposite end of the tray. The reason is head loss of the liquid as it flows across the bubbling area of the tray deck. As you can see from Figure 9-4, the vapor flows preferentially to the path of least resistance: that is, where the depth of the liquid is least. This results in vapor–liquid channeling. But, of greater importance, the high localized vapor velocities promote localized liquid entrainment. Both the channeling and the entrainment will reduce tray fractionation efficiency.

Picket weirs (see Figure 9-5) mitigate these problems by increasing the weir loading. As the weir loading is increased, the crest height is increased. As the crest

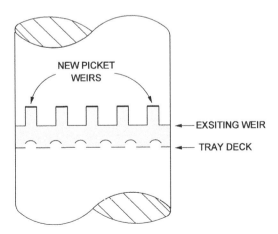

**Figure 9-5**   *Picket weirs improve fractionation efficiency at low liquid rates.*

height is increased, the depth of liquid on the tray deck rises. The increased depth of liquid on the tray deck improves the tray efficiency in several ways:

- The liquid flows more evenly across the tray, and stagnant pools of liquid are reduced. The picket weir acts like a restriction orifice distributor where the orifices are the space between the pickets.
- The overall deeper liquid level on the tray reduces the relative difference between the depth of liquid on the two sides of the tray. This reduces both vapor–liquid channeling and droplet entrainment, due to high localized vapor flow.

The reader may well ask how I come to know these things. Actually, I don't. These ideas are the conventional wisdom among tray designers. But what I am certain of is that at low liquid rates, picket weirs positively improve fractionation efficiency without generating more internal reflux. I have retrofitted enough towers with picket weirs to be 100% sure of this fact, based on the actual before and after plant lab distillation data.

## PICKET WEIR DESIGN

The picket weir height should be half the tray spacing. If the weir loading is about 4 gpm/in., I do not use picket weirs. If the weir loading is less than 1 gpm/in., I consider picket weirs to be critical. The space between the pickets should be such that the weir loading is increased to about 4 gpm/in. The width of each picket is roughly 5 to 8 % of the tower diameter. For two-pass trays, picket weirs may be required at the center downcomers but not be needed at the side downcomers. Notched weirs are as good or better than picket weirs but are not generally used on perforated fractionation trays.

## INCREASING INTERNAL REFLUX

If I wanted to improve fractionation between the light distillate and the heavy distillate shown in Figure 9-3, I could increase the flow of liquid on the fractionation trays between the two pump-arounds. To increase the internal reflux rate on trays 13 to 16, I would:

- Reduce steam production in the heavy distillate pump-around section.
- Increase the heat extracted to cooling water in the light distillate pump-around section.

The higher liquid flow on the fractionation trays might eliminate the need for picket weirs. But this would entail a reduction of steam production in the heavy distillate pump-around. The lost energy of the steam would appear as heat loss to cooling water in the light distillate pump-around. Thus, in a broader context, picket weirs can be viewed as an energy-saving device. If we do not make steam in the heavy distillate pump-around cooler (i.e., steam generator), the steam will have to be produced in the power station by burning fuel. This then produces $CO_2$ and nitrogen oxides, both of which contribute to global warming.

Picket weirs are an example of how the process engineer can substitute knowledge for energy consumption. Isn't that what process engineering is all about?

## PRESSURE OPTIMIZATION

Reducing a fractionator operating pressure, if not limited by entrainment, is the most energy-efficient manner of improving component separation. Increased volumetric flow of vapor will improve vapor–liquid mixing. Less tray deck leakage resulting from the greater vapor volume will also help. In addition, the relative volatility of key components always increases as the tower operating pressure is reduced. With better fractionation, both the reflux rate and the reboiler duty can be reduced and still meet product specs. Further energy savings will result from the lower tower bottom temperature as liquids boil at a lower temperature and a lower pressure.

However, if the lower pressure and greater vapor velocities cause entrainment and jet flooding, reduction in pressure is counterproductive. At a fixed reflux rate, the tower operating pressure is slowly changed. The degree of product separation will then indicate the optimum fractionator operating pressure. At lower charge rates, the optimum tower operating pressure will normally be less than at higher fractionator feed rates. Of course, tower pressures cannot be reduced when the tower operation is limited by condenser capacity.

# Chapter 10

# Increasing Centrifugal Pump Capacity and Efficiency

It was raining hard. My coveralls were saturated, my boots were saturated, and I was saturated. I stood dripping in the center of the control room floor. Maybe 44 years of this pressure drop survey stuff is enough? The cold damp had gotten into my knees and my hands.

The crude tower bottom stripping trays were fine; their $\Delta P$ value was normal. The reason for the high diesel oil content of the crude fractionator bottoms was not stripping tray fractionation efficiency. It was just that the flash zone pressure was 20 psi too high. The high flash zone pressure was suppressing the vaporization of the diesel oil boiling-range components. Diesel was slipping out the bottom of the crude tower and into the vacuum tower feed. The LVGO (light vacuum gas oil) was 60%, 650°F and lighter. The LVGO was part of the FCU feed. Each barrel of diesel charged to the cat cracker (FCU) lost the Shalom refinery in Texas $12 to $15. But I was too tired and too cold to care.

Suddenly, a man in a shiny white hard hat, blue jacket, and red tie rushed up to me. "Mr. Lieberman, I presume? It's so nice to meet you. I have heard so many wonderful things about you." He reached for my wet and very dirty hand.

"Sorry, my hands are dirty," I mumbled.

"Oh! That's fine. I've heard so much about you. I'm such a big fan of yours. My mother talks about you all the time."

"Your mother?" I was surprised. Even my own mother didn't particularly like me all that much.

---

*Process Engineering for a Small Planet: How to Reuse, Re-Purpose, and Retrofit Existing Process Equipment,* By Norman P. Lieberman
Copyright © 2010 John Wiley & Sons, Inc.

"Yes, my mom. I'm Peter Rabowski, the plant manager here at Big Springs. We really appreciate your helping us with our crude pipe still operation. I know you must be very busy. Thanks for coming."

"If you've got the money, I've got the time. But Mr. Rabowski, have we met before? You look kind of familiar."

"No, Norm. I haven't had the pleasure until now. But you knew my mom. Lydia Rabowski. She worked with you at Whiting, on the FCU."

"Yeah. Does Lydia still work at the plant? I remember her. She has long blonde hair and. . . ."

Peter laughed, "No, Norm. Mom's 74 years old. She broke her hip last May. Been in a wheelchair since. Her hair went gray when I was still in high school. But she remembers you. But look, this diesel oil in cat feed's a big problem. We have 14% 650°F and lighter in the crude fractionator bottoms. I read in your book *Troubleshooting Refining Processes* that it should be like 5% 650°F and lighter."

"Where did you get that edition of my book? It was published by Pennwell in 1980," I asked, quite surprised.

"Oh, mom had it. She used to read it every night while I did homework. She has it still."

"Maybe I knew your dad, too. What kind of work did he. . . ."

"Gee, I have a 10:00 A.M. teleconference with the V.P. in Paris," Peter interrupted. "Look, Norm, let's figure out how to get that 650°F minus stuff out of the LVGO. If nothing else, we could build a diesel oil recovery tower."

"A what?" I asked. But Peter Rabowski had rushed off.

## HYDRAULIC LIMITATIONS

The crude fractionator flash zone pressure (see Figure 9-2) was running at 50 psig because the tower overhead receiver pressure (Figure 10-1) was operating at 40 psig. The operators on the crude unit had their pressure control valve (P-1) set for 30 psig in an attempt to minimize the crude fractionator pressure. Hence, the PRC valve for P-1 shown in Figure 10-1 was wide open.

The wet gas compressor shown in Figure 10-1 served the entire refinery. It was operated by a remote group of operators. Remote in an intellectual sense. They all worked in the same central control room. They all went duck and deer hunting together. As far as their panel boards were concerned though, they operated in parallel universes.

The problem was the condensate drum level. The LRC (level recorder control valve) was 100% open. Should the level in this drum carry over, liquid would wreck the wet gas compressor. Thus, the wet gas compressor operators raised the condensate drum pressure PRC (pressure recorder control) set point to 38 psig. This caused the wet gas compressor spill-back valve to open.

I then noted the following temperatures (using my infrared temperature gun) in the field (see Figure 10-1):

- Condensate drum temperature = 80°F
- Compressor discharge temperature = 180°F
- Compressor suction temperature = 140°F

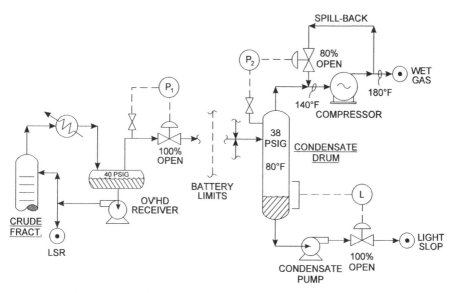

**Figure 10-1** *Condensate pump limits diesel recovery from crude.*

Using the *lever rule* I calculated that 60% of the gas flow to the compressor was recirculated wet gas. As the compressor was a motor-driven (i.e., constant-speed) reciprocating machine (i.e., constant suction volume), I calculated that I could reduce the suction pressure from 38 psig to 7 psig, as follows:

*Step 1.* 38 psig = 52 psia (i.e., 14 psia atmospheric pressure in Big Springs)
*Step 2.* 52 psia × 40% = 21 psia
*Step 3.* 21 psia – 14 psia = 7 psig

The 40% represents the percent of gas flow to the compressor not recirculated through the spill-back valve, $P_2$.

## AN OLYMPIC CHAMPION

Josh is the tech service manager at the Shalom refinery. His son was the quarterback for the Hook-em Horns high school football team.

"Josh," I explained, "We can lower the crude tower pressure if we can solve the condensate pump head problem. It's a matter of pump hydraulics."

"Norman, my good friend. Do me the honor of coming to the Hook-em Horns pep rally tonight. I have approval from Mr. Rabowski to order a new condensate pump. We agree it's needed."

"A new pump? What's wrong with the existing pump?"

"Sadly, my good friend Norman, the existing pump is too old. It fails to develop enough discharge pressure unless we raise the condensate drum set point pressure to 38 psig. It's very sad."

"But, Josh, you and I are both older than that pump. We're still okay."

"The maintenance department has overhauled this pump many, many times. They have tried very, very hard. But the pump is tired and old. Norman, my good friend, I too am getting old. It's very, very sad."

"So where's the new pump?"

"Ah, my friend. We have yet to go out for bids. The funds for the new pump are in next year's budget. Sadly, we lack funds this quarter."

"Next year?" I asked.

"Yes, Norman. But perhaps you will wish to visit the spare condensate pump. It's in the maintenance shop. The maintenance foreman, John Brundrett, will also be at the rally tonight."

## WORN IMPELLER-TO-CASE CLEARANCE

I never accept the idea that centrifugal pumps cannot run on their performance curve provided that the available NPSH (net positive suction head) is adequate. My first step was then to measure the efficiency of the conversion of the motor's energy output to pumping energy. To make this measurement, I needed the following data obtained from the pump:

- Pump inlet temperature = 85°F
- Pump discharge temperature = 90°F
- Pump flow = 50 gpm
- Fluid specific gravity = 0.60
- Pump suction pressure = 40 psig
- Pump discharge pressure = 200 psig
- Specific heat of liquid = 0.5 Btu/(lb-°F)

With these data I calculated the percent of the motor's energy going to heat:

1. Calculate the pounds per hour of liquid flow and heat increase as follows:

   - (50 gpm) (60 min/hr) (8.34 lb/gal) (0.60 S.G.) = 15,000 lb/hr

   - Heat = (15,000 lb/hr) (0.5 sp. heat) (90°F − 85°F) = 37,000 Btu/hr

2. Calculate the energy going to work in increasing the fluid pressure. Note that there are 775 ft-lb in 1 Btu. Obtain the feet of head:

   - $(200 \text{ psig} - 40 \text{ psig}) \left( \dfrac{2.31 \text{ ft}}{\text{psi for water}} \right) \left( \dfrac{1}{0.60 \text{ S.G.}} \right) = 618 \text{ ft}$

   - $(618 \text{ ft}) (15,000 \text{ lb/hr}) \left( \dfrac{1}{775} \text{ ft-lb/Btu} \right) = 13,000 \text{ Btu/hr}$

The percentage of motor energy going to parasitic heat losses was a staggering

$$\bullet \quad \frac{37,000}{(37,000 + 13,000)} = 74\%$$

This was more than double the 30% typically lost to heat by a small centrifugal pump. The pump was certainly mechanically defective. The defect was causing internal liquid recirculation. That is, for every gallon of liquid leaving the pump, 2 to 3 gallons was being recirculated internally. This internal recirculation could be caused by only two problems:

- The impeller wear ring was worn out. The eye of the impeller spins inside this replaceable ring. The ring is easily replaced by a new factory-supplied part.
- The clearance between the outside diameter of the impeller and the inside of the pump case has increased. Either the impeller is worn down, or the interior of the pump case is worn with erosion and corrosion. Replacement of the impeller is easy with a new factory-supplied part. Repair of the pump case is difficult and time consuming, as explained below.

## VISIT TO THE PUMP REPAIR BENCH

John Brundrett, the maintenance chief, was reading the sports page of the *Hook-em Horns News* when I dropped in.

"Mr. Brundrett? Norm Lieberman. Glad to meet you, sir," I said.

John eyed me with suspicion, "So, what do you need?"

"Well, John, I'd like to inspect P-74, the wet gas compressor condensate pump."

"You just passed it. It's on the bench. Go look," John Brundrett picked up his paper and resumed reading.

"Sorry, John, but I want to inspect its internals," I said.

"Say. Who are you? What's your name again?" I was making progress, John had put down his newspaper. "We just put in a new 317 stainless steel impeller wear ring and a new full-size $8\frac{1}{2}$-in. chrome impeller. It's all new. What did you say your name was?"

"Lieberman. Pete Rabowski told me to check the pump case for internal wear. Check the impeller-to-case clearances. I used to work with Pete's mom in Indiana," I said lumping fact and fiction together.

John Brundrett rose; I had ruined his peaceful afternoon. "Dave," he screamed above the machine shop noise, "tear down that [expletive deleted by publisher] P-74 again."

"It's an old pump. Should really be replaced. Just look at the interior of the case. It's all chewed up. Corrosion and erosion," John observed. "I guess you're right, Lieberman. I tried to order a new pump case. But the manufacturer went bottoms-up in 1955. Yeah. The clearances between the case and impeller are 10 times too big.

A definite cause of internal recirculation and lost head. Wonder how it pumped at all, it's so old."

"You know, John. I've seen this before, when I worked in the Brooklyn Navy Yard in 1962. You need to build the interior of the pump case back up with layers of welding rod metal. Then you machine it back down on a metal lathe to restore the original tolerances."

"Sure. We could do that. Dave could sure do that stuff," John observed, "But that's an awful lot of work."

"Oh, but I thought you all got paid to do that," I said.

"Paid for what?" Dave asked.

"Work," I answered.

## THE PEP RALLY

John Brundrett was too busy to talk to me at the rally. But I had a great time telling Lydia about how I saved buying the new condensate pump.

"Just like the old diesel oil recovery tower project in Whiting," she laughed. "Smart engineering is worth more than gold. Saves time, energy, and steel. I tell Peter that all the time. 'Use what you've got. Don't buy new equipment. We live on a small planet. Our resources are limited. The Creator would not look kindly on a new light condensate pump.'" Lydia waved to her grandson down on the football field. "You know, Norm, I do miss those old days in Whiting."

Two days later the rebuilt condensate pump went into service. Dave came down from his machine shop to watch it being run-in. The operators cranked down on the $P_2$ pressure set point (Figure 10-1). As the condensate drum pressure fell to 10 psig, the crude fractionator overhead receiver pressure dropped from 40 psig to 14 psig. The crude fractionator flash zone pressure also fell from 50 psig to about 30 psig.

The sample of fractionator bottoms taken on the following shift showed an ASTM D-86 distillation analysis of 6% for the 650°F minus components. The diesel oil in LVGO dropped to about 20%. Pete Rabowski and Josh were both pleased. However, Dave, the machinist, was even happier.

"Hey, Lieberman. Get those guys to send me that old spare P-74. No sense buying new pumps when we can fix up the old ones. My old man would have a fit if he saw the stuff we waste in this plant."

## OTHER METHODS TO INCREASE PUMP CAPACITY

I don't want to discount the importance of the impeller wear ring. This pump component needs to be kept in good mechanical condition. A worn-out impeller will be indicated by a reduction in the amperage load on the motor driver.

Certainly, increasing the size of the impeller is a normal way to increase pump capacity and head. I always keep the maximum impeller size to not more than $\frac{1}{4}$ in. of the maximum. The amp load on the motor driver will increase with the diameter of the impeller cubed (power of 3). That is, going from a 10- to an 11-in. impeller will increase the motor amperage by 34%. The FLA (full-limit amp) load on the motor driver should not exceed the increased motor amp load from the larger impeller.

Reducing the downstream control valve $\Delta P$ value will also increase the pump capacity. To check if the $\Delta P$ on the control valve is a problem, first open the control valve 100%. Then fully open the bypass around the control valve. If the flow increases by more than 5%, the control valve size, or internal trim, should be increased.

Calculating the control valve pressure drop does not tell us how much pump discharge pressure is lost. The reason is, as a liquid leaves a control valve, it slows down. Thus, some portion of the control valve $\Delta P$ is recovered due to the conversion of velocity partly back into pressure.

Check the $\Delta P$ value of downstream flow orifice plates. Some orifice plates are sized for 400 in. $H_2O$ (14.4 psi) $\Delta P$. That is okay if you are not limited by the pump capacity. The problem is that some but not all pressure loss through the orifice plate is recovered. Changing to an orifice plate that is sized for 100 in. $H_2O$ $\Delta P$ will eliminate most of this head loss (i.e., increase the orifice plate hole diameter by 41.4%).

Strange to say, but pump capacity can actually be increased by reducing the impeller size! This happens if the pump is limited by the size of the motor. To keep the motor from tripping on excess amperage load, the operators will keep the pump discharge valve throttled. The smaller impeller will permit the discharge valve to be left open, thus saving energy as well as capacity.

Oversizing motors does not waste a significant amount of energy. As a design engineer, I will specify a motor for a new pump in accordance with the maximum impeller size rather than the current impeller size. Only about 3% of the unneeded motor horsepower is wasted, due to the motor efficiency factor. But then a larger motor and wiring will not be needed when eventually a larger impeller is required for the pump.

For steam turbine–driven pumps, always check the pump speed first. If the speed is less than rated, the problem lies not with the pump but with the steam turbine horsepower output. For example, is the turbine steam supply regulator governor valve fully open? Check to see if the hand or horsepower valves are all open. Opening a single port by opening a hand valve will typically increase the turbine power output by 10%. Also check if the governor valve on the steam supply line is running wide open.

## MARGINAL CAVITATION

Large high-head centrifugal pumps that operate with low suction pressure have a distinct tendency to drop-off their operating pump curve when deprived of adequate suction pressure or NPSH. The pump discharge pressure is quite steady, as is the discharge flow. Also, the pump does not make any unusual cavitating type sounds or vibrations. To all external appearance, the centrifugal pump is running normally. But if you increase the suction pressure by a few psi, its discharge pressure might increase by 100 psi. If you now increase the pump's suction pressure by a few more psi, the pump discharge pressure would only increase by the same few psi. I have seen this odd behavior in:

- Coker fractionator heavy gas oil circulating pumps
- Crude pre-flash tower bottoms pumps
- Vacuum tower pump-around service

It seems that such centrifugal pumps run on an inferior operating curve when slightly short of NPSH. But they regain their normal operating curve when adequate suction head is provided. What also seems strange is that although running in a suction pressure limited mode, these pumps do not have a history of excessive failure of their mechanical seals. Hence, the capacity of such pumps may be increased simply by raising the level in the feed vessel by several feet or by subcooling the liquid slightly.

## EFFECT OF TEMPERATURE

Reducing the operating temperature of the pumped liquid will have the following effects:

- *For hydrocarbons:* Each reduction in temperature of 100°F will increase the specific gravity by 5%.
- *For aqueous systems:* Each reduction in temperature of 100°F will increase the specific gravity by 1%.

Hence, for aqueous systems, the effect of temperature on centrifugal pump performance can be neglected except for NPSH effects. For hydrocarbons, operating in a nonviscous range [below 40 centistoke (cS) or centipoise (cP], the effect on a centrifugal pump is significant for two reasons:

- Although the volumetric flow may be the same, the mass of liquid moved will increase in direct proportion to the increased specific gravity.
- Of greater importance is the effect of specific gravity on the pump's discharge pressure. A centrifugal pump operating at a fixed volumetric flow rate at a constant speed, pumping a fluid of low viscosity, produces the same feet of head regardless of the specific gravity of the fluid. For example, at the Cenneco Oil refinery in Louisiana, I had a hydrocracker charge pump moving 18,000 bsd with a pump discharge pressure of 2600 psig. I cooled that liquid off by 85°F and the discharge pressure increased to about 2700 psig. With the extra pump discharge pressure, I was able to increase the hydrocracker feed rate to 22,000 bsd. Note that the hydrocracker reactor pressure was about 2400 psig.

## EFFECT OF VISCOSITY

Viscosity does not appear to have much of an effect below 100 cS (500 SSU) on smaller (50 hp) pumps. On larger pumps, high viscosity (above 50 cS) is a killer to both head and flow. To overcome this problem, I once mistakenly changed out a crude charge pump (60,000 bsd, 200 psi differential pressure), to an expensive positive-displacement gear-type pump. Positive-displacement pumps will tolerate a much higher viscosity range than will centrifugal pumps.

I had replaced one of the two existing centrifugal pumps with a positive-displacement pump because my client, the Coastal Corporation in Corpus Christi, Texas, was planning to run a very heavy, high-sulfur, high-viscosity crude from Venezuela. When the new heavy crude was received, my new gear pump handled the heavy crude without a problem. But so did the old centrifugal pump that Coastal had kept as a spare!

What had I missed? It isn't that the published data for viscosity vs. pump efficiency is inaccurate. It's that the crude viscosity was lower than the published crude analysis from Venezuela. The discrepancy was in the light ends dissolved in the crude:

- Methane
- Ethane
- Hydrogen sulfide

Due to losses on handling in the analytical laboratory, crude analysis always understates these light ends. While the amount that is lost in the lab is tiny (perhaps 0.1 wt% on crude), the effect on viscosity can be pronounced.

I noted the same effect on a hot coker feed tank designed for Tenneco Oil in 1986. The 50,000-barrel tank was to be operated at 500 to 550°F. I knew that the 5% sulfur resid would evolve hydrogen sulfide, due to thermal cracking. Even though the lab analysis indicated a viscosity requiring a positive-displacement pump, I used an ordinary centrifugal pump, which performed just fine.

If the capacity of a centrifugal pump (head or flow) is limited by high viscosity, a few percent of a light diluent can eliminate the problem. Alternatively, a moderate increase in pumping temperature can cut viscosity in half for hydrocarbons. The same is true for pumps limited by the driver horsepower. If a motor is tripping-off on high amps, reducing the viscosity of the pumped fluid from 60 cS to 30 cS will typically eliminate the need for a larger motor driver.

## NPSH-LIMITED PUMPS

In 1983, the old No. 1 vacuum tower at the Poseidon refinery near New Orleans was revamped. The process limit of the tower was known to be a maximum pump-around rate of 12,000 bsd. I submitted a bid to provide the process design for the "de-bottlenecking" project for $20,000. A competitive design company submitted their bid for $200,000. Consistent with a long history of corruption in south Louisiana, my competitor received the contract.

To increase the capacity of the pump-around circuit, a new 20,000-bsd-rated centrifugal pump was installed. On startup, the capacity of the new pump was found to be only 12,000 bsd. What had happened?

Gerry Harris, the operating manager, asked me to look into the pumping deficiency. I reported the following to Mr. Harris:

- The draw-off nozzle from the tower is only 4 in. in diameter.
- At 12,000-bsd flow, the head loss through the draw-off nozzle equals the height of the internal draw-off sump:

$$\Delta H \text{ (in.)} = 0.34 \text{ (velocity, ft/sec)}^2$$

- Pump-around rates above the 12,000-bsd limit will just cause liquid internal overflow and loss of the pump suction pressure.

To prove my point, I dragged Gerry out of his air-conditioned and paneled office and showed him the pressure gauge I had placed at the suction of the new pump. At 13,000 bsd, the pump suction pressure disappeared in 3 minutes.

"See, Gerry," I explained, "This pump is limited by NPSH. The draw-off nozzle is too small."

"But, Norm, the new pump has a new 6-in. suction line rated for the full 20,000-bsd design flow," Gerry protested.

"Gerry! It's the 4-in. draw-off nozzle at the tower that's limiting the flow, not the suction line or the pump! This is all a waste. There was nothing wrong with the old pump," I screamed above the roar of the vacuum heater burners.

"So, Norm, submit a competitive bid to correct the problem," Gerry answered.

So I submitted a bid for $20,000 for a new process design to increase the capacity of the No.1 vacuum tower. And the same competitor design company with an expensive office in the French Quarter submitted a bid for $200,000. Being New Orleans, you can guess who got the contract.

# Chapter 11

# *Eliminating Process Control Valves Using Variable-Speed Drivers*

My oldest and very excitable daughter, Lisa, rushed into my office last month. "Dad, I'm trading in my SUV for a hybrid. It will save a ton of gas. I'm tired of those high gas prices."

"Lisa, your SUV is new; the trade-in value will be less than half what you just paid."

"I knew you'd say that, dad. You're just sticking up for those rip-off oil companies again."

"Lisa, the hybrid won't save you much gas. You do most of your driving on I-10. Anyway, 75% of the cost of a gallon of gas is due to crude oil and 10% is taxes. Refinery margins are really small."

"Always sticking up for those crooks in the oil business. Dad, why can't you invent something like a hybrid car machine? You know," Lisa searched the ceiling for her idea, "like something they could use to make gas a lot cheaper. Don't your refineries waste half the crude oil just making the gas?"

"No, Lisa. Seven percent. Not half."

## VARIABLE-SPEED ELECTRIC MOTORS

Actually, we do have something in the process industry akin to a hybrid car. It's something similar to the idea of converting a car's momentum into stored electric

*Process Engineering for a Small Planet: How to Reuse, Re-Purpose, and Retrofit Existing Process Equipment.* By Norman P. Lieberman
Copyright © 2010 John Wiley & Sons, Inc.

potential. The idea is older than control valves, older then the process industry itself. It is to slow down equipment when less work is required. The concept of converting work to friction when we are working too fast is a strange, new, and antisocial idea. But that's what control valves on centrifugal pump discharge lines are doing: destroying electric work—often, more than half the work that the electric motor has just produced at enormous cost.

The problem originated with Nicoli Tesla, who invented:

- Radio (and the basis for cell phones and TV)
- Fluorescent lighting
- Induction heating
- Alternating-current (ac) polyphase motors
- Radar
- Remote control
- The high-voltage ac electric distribution system used all over the world
- Advanced turbine design

Tesla's design of the ac motor, presented to the world in the 1880s, based on a rotating magnetic field, meant that the motor speed was a function of the frequency of the electric current. Since our electric grid has a fixed number of cycles per second (60 in the United States, 50 in Europe), it follows that:

- U.S. single phase: 1800 rpm
- U.S. three phase: 3600 rpm
- European single phase: 1500 rpm
- European three phase: 3000 rpm

To change an ac motor speed, you can switch the wiring of the motor to change its number of phases or change the electric current supply frequency (cycles per second). It is entirely possible, and has been since Tesla's first motor, to alter the speed of the motor in continuous small increments by changing the frequency (cycles per second) of the electric current. As the electronics industry has advanced, this has become progressively more practical and less costly.

## ELIMINATING CONTROL VALVES ON A PUMP DISCHARGE

Eliminating control valves is not a new idea. I have seen this used in practice at a 30-year-old delayed coker HCGO pump-around at a refinery in California. Figure 11-1 shows the concept. The frequency of the electric current is controlled by the flowmeter. The current's frequency then controls the motor and pump speed. The parasitic energy

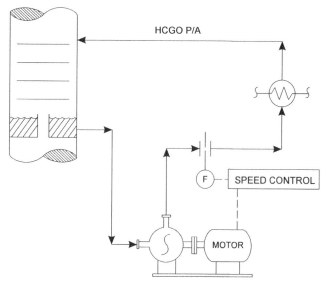

**Figure 11-1**   *Variable-speed motor controls pump around flow without a control valve.*

lost through the control valve, along with the valve itself, is gone. Of course, there are several potential problems.

First, unless precautions are taken, the variable-frequency transmission between the switch gear room (substation or breaker building) and the motor will create radio waves that can interfere with the internal plant radio communication systems. This is a minor problem if addressed beforehand.

It is interesting how, in Tesla's mind, the radio, the ac motor, induction heating, and remote control were all aspects of the same property of nature: transmission of energy through the "ether." Tesla died in 1943 in New York City, just a few months after and a few miles from where I was born. By then, the existence of the ether had been disproved.

Second, not all motors are created equal. Cheaper motors employ aluminum components. Such motors will overheat if used in a variable-frequency service. So if the intent is to modify an existing service for speed control as shown in Figure 11-1, the cost of a new motor may possibly have to be included in the project. I would guess that a suitable 3600-rpm 460-V 100-hp three-phase totally enclosed fan-cooled induction motor might cost $12,000 (2008 prices in U.S. dollars).

The electronic hardware required to vary the frequency to the motor might cost a similar amount. By the time the installation is completed, a knowledgeable rotating equipment expert engineer (not me) has estimated a cost of $50,000 to $100,000. With energy and electricity prices rising, I would guess that electric power might be valued at 20 cents a kilowatt, or 16 cents per horsepower, within a few years. This means that at the very most, a 100-hp driver would consume $140,000 per year of electric power.

Just like a hybrid car, I suppose that one might reasonably expect to save about 10% of the maximum power use. Why only 10%? There are several reasons:

- The motor itself will be used at only some fraction of its rated horsepower of 100.
- The pump is spared; and sometimes the spare pump is running.
- If the control valve $\Delta P$ value was always a large percentage of the pump-developed $\Delta P$, the more economical approach would be to reduce the centrifugal pump impeller size (see Chapter 10), to keep the control valve in a more open but still controllable position.
- Finally, the process plant itself does not run all the time.

So if even at the higher projected cost for power, it rather looks like the payout for the retrofit I have described will be about five to seven years. At no refinery at which I have ever worked could such a nebulous incentive with a five- to seven-year capital recovery period result in the project being approved by management.

It's the same argument I advanced with my daughter Lisa, who rushed to trade in her SUV for the hybrid car. Okay. But there's another way to look at both situations— That is, from the perspective of one who wishes to help save the planet from the ravagers of humankind.

What's so bad about a five- to seven-year period for capital recovery investment? I certainly would not like to support such a project payout to increase crude distillation capacity, or to build ethanol plants, or to convert natural gas to distillates. But this is different. A variable-speed motor driving a centrifugal pump is sure to save electric power and reduce greenhouse gas emissions. How much and how fast is hard to know. The main thing is that we are going to eliminate the parasitic control valve losses forever. We are moving in the right direction. We are on the right path. Responsible management decisions should be guided by this criterion as well as by profitability.

## NEW UNIT DESIGN

For a grass-roots project, the economics are quite different. Certainly, the cost of a new motor is not a factor, as a motor will be purchased regardless. Also, the cost of the control valve will be saved. The cost of the two isolation gate valves to block in the control valve, as well as the bypass line and its gate valve, will be eliminated. Finally, the expense of the control loop to and from the control valve will also be saved.

I would guess that the net cost difference between having a parasitic control valve on the discharge line of a pump will more or less equal the cost of the electronics hardware to control the pump flow, or other process variables, with a variable-frequency motor drive.

## STEAM TURBINES

This discussion can easily be extended to any steam turbine–driven centrifugal pump. These machines have always been variable speed, the speed being controlled by the "governor valve" on the motive steam inlet. Why $99\frac{1}{2}\%$ of steam turbine–driven pumps are equipped with parasitic control valves on their discharge is a total mystery to me. The few installations that I have seen, where the speed control governor valve is controlled directly by a process variable (flow or level), work just fine. No control valve exists, and none are needed. Whereas retrofitting a centrifugal pump's ac motor for variable speed is expensive and somewhat complex, direct process variable control without a control valve for a steam turbine–driven pump could not be any simpler.

Ladies and gentlemen, fellow passengers on this little green planet. The use of a process parameter to vary the speed of a steam turbine directly, without a control valve is really ancient technology. It really ought to be a standard process engineering design practice.

So Lisa didn't buy a new hybrid. But she is still convinced that refineries are run by crooks and that I am aiding and abetting a criminal conspiracy. Maybe she's right. But dear Lisa, I'm trying to do better. I'm looking for the right path. I can see the light. It is a long way off, but I can surely see it.

## VARIABLE-SPEED COMPRESSORS

Reciprocating compressors could profit from variable-speed drives at reduced loads. However, there is a simpler and quite energy-efficient method of unloading a reciprocating compressor called an adjustable head-end unloading pocket (see Chapter 15). The only disadvantage of the head-end unloader is that it's expensive. Also, it can only reduce the capacity of the reciprocating compressor by 25% maximum. All reciprocating compressors can easily be retrofitted (but not cheaply) with such an unloader.

Centrifugal compressors are very often purchased with variable-speed gas turbine or steam turbine drives. Many others are driven by fixed-speed motors and then have their suction pressure controlled by either:

- A discharge-to-suction spill-back line, also called an antisurge line, or
- A suction throttle valve, typically used to directly control the process upstream pressure.

The complication in using variable-speed drives on centrifugal compressors that have not been designed for such applications is that one needs to guard against "surge" and "stonewalling." To know when these problems would be encountered, we would need the compressor curves at various operating speeds, which may not be readily available for a compressor intended for a fixed-speed drive.

The stonewalling problem just determines the maximum gas flow that the compressor can achieve. It is an operating issue that could be overcome by running the centrifugal compressor faster. Stonewalling will not damage the compressor.

Surge is another matter. Running a centrifugal compressor in a surging mode for very long will cause pieces of the rotor to fly through the compressor case and kill someone. I do not mean to suggest that variable-speed motors are not applicable to almost all centrifugal compressors. I am sure that they are and that huge electric power savings are potentially obtainable. It is just that evaluation must always be on a case-by-case basis, with the vendor compressor performance curves at hand, for a range of speeds, and after consultation with the vendor.

As I write these words, I have a mental picture of the 12,000-hp motor driving the wet gas compressor at the Jupiter refinery's delayed coker unit in Mississippi (see Chapter 1). The suction throttle valve was mostly closed and I estimated that several thousand horsepower were being lost due to suction throttling of the wet gas. Here, I thought, would be a nice application for a frequency-adjusted variable-speed electric motor.

In Chapter 15 I review the relationship of a centrifugal compressor when varying the following parameters:

- Molecular weight
- Number of wheels
- Suction pressure
- Surge
- Polytrophic head
- Compressor speed

For now, I will simply note that the problem of variable-speed operation is more complex for a centrifugal compressor than for a centrifugal pump.

## SPILL-BACKS WASTE ENERGY

The three-way valve shown in Figure 11-2 is a real source of energy waste. In a sense it's the exact opposite of using variable speed to control a process flow. From safety and reliability, the three-way valve is a positive factor. Typically, the valve can be adjusted to maintain a maximum pump discharge pressure or a minimum pump flow. It's really dangerous to run a large high-head pump much below 60% of its best efficiency point. The pump's internal components, including the mechanical seal, may well be damaged.

But the liquid recirculated back to the pump suction represents quite a bit of wasted energy. Setting the three-way Yarway valve to maintain a lower flow or a higher discharge pressure would clearly reduce motor horsepower, but at the expense of running the pump farther back on its operating performance curve, and perhaps compromising the integrity of the mechanical seal.

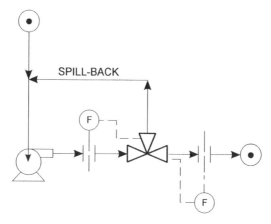

**Figure 11-2** *Three-way valve is an example of wasting power by using spill-back control.*

## IMPELLER DOWNSIZING

One way to mitigate the effects of running with the three-way valve spill-back too far open, or operating with the discharge control valve too far closed, is to reduce the size of the pump impeller (as discussed in Chapter 10). This is a cheap alternative to gaining part of the benefit of speed control. The objective is to reduce the size of the impeller so as to keep the control valve in a mostly open, but still controllable position.

As a young engineer, it did not appear to me that many control valves were running mostly closed. But now I know better. To keep many control valves in a reasonably controllable position, two measures are taken:

- Operators partly close the gate valves at the pump discharge. Obviously, this wastes hydraulic horsepower. But also, the seat in the gate valve will be eroded with time. Then, when the pump must be isolated for repair, the gate valve will leak through.
- Instrument engineers undersize the control valve. In Lithuania I once saw a 1-in. control valve in a 4-in. process line! But everyday we see a 6-in. control valve in an 8-in. line. I know very well that it's standard practice to have a control station one line size smaller than the process line. But who set the standard? Not someone interested in saving driver horsepower and motor amps.

## ESTIMATING INCENTIVE FOR VARIABLE-SPEED DRIVES

For turbine-driven pumps, no such estimate is required. The process variable should be linked to the governor, not to a parasitic control valve. All turbine manufacturers support this practice.

For motor-driven pumps, I would proceed as follows:

*Step 1.* Measure the $\Delta P$ developed by the pump.

*Step 2.* Measure the $\Delta P$ across downstream heat exchangers, vessels, and piping, but not control valves.

*Step 3.* Step 2, divided by step 1, is the useful percent of the pump's energy. The remainder is parasitic control losses.

*Step 4.* Use a clamp-on amperage meter to measure the current amperage consumption of the motor.

*Step 5.* Multiply the parasitic energy loss in step 3 by the amperage measured in step 4. This is the electric power that can be saved by variable motor speed control.

Variable-speed motors will also reduce the heat load on motor windings. All motor windings must eventually be replaced. Often, it's easier to discard a motor than to have the coils rewound. But running the motor at a minimum load will keep the coil windings much cooler and thus they will last far longer.

## THROTTLING MOTIVE FORCE

In a general way, the concept that I'm promoting in this chapter is to avoid controlling a process parameter in a parasitic manner. It's best to throttle-back (or slow down) the motive force driving the equipment rather than applying a brake to the equipment. After all, you wouldn't apply constant pressure on your car's gas pedal with your right foot and then control the car's speed while braking with your left foot.

For example, when possible, it's best to control the motive steam pressure to a vacuum steam jet rather than to spill-back the tail gas from the jet's discharge-to-suction to maintain a constant vacuum tower top pressure. Admittedly, due to the nonlinear nature of vacuum jets, it's simpler to control tower pressure on spill-back rather than throttling the motive steam pressure to the jets. But that's a price we will have to pay because we live on a small planet with 7 billion other human inhabitants.

Extrapolating this sort of logic further: If I have a distillation tower that works best at a lower pressure (as most towers will, due to a greater relative volatility at lower pressures), should I not always minimize the tower pressure? The lower pressure will allow me to save energy by running the tower cooler and with less reflux (a beneficial consequence of the greater relative volatility). I could simply allow the tower pressure to drop, as the facility's cooling-water temperature declines each evening. Or, allow the tower pressure to slip down, as the ambient conditions make my finned air-cooled overhead condensers work better in the winter. Of course, now I will have to adjust my target temperatures continuously as a function of the slowly varying distillation tower pressure, to keep my products on-spec.

I call this "floating tower pressure control"; it is practiced informally by many operators running refinery crude distillation towers that are limited by fired heater capacity. This concept of floating tower pressure control, would, of course, work best when linked with a computer to adjust the tower temperature targets as a function of the tower operating pressure. Or one could dispense with temperature targets altogether and use an online gas chromatograph to keep products on-spec, by changing the reflux and reboiler duties, as the tower pressure slowly follows changes in ambient conditions.

# Chapter 12

# *Expanding Refrigeration Capacity*

I had gone down to pray by the banks of the Mississippi as it flows under the Huey P. Long Bridge. Guided by tugs, barges loaded with wheat and soybeans flowed downriver to feed the world. Struggling against the current, tankers from the Urals, Nigeria, and Basra pushed upriver. In the 1960s we traded crude for food. Now we trade crude for promises. Promises to pay that we will not keep. Over 1.2 trillion pounds per year of hydrocarbons from the frozen sands of northern Alberta to the humid forests of Colombia rush into Louisiana, California, and Texas to feed the ravenous refineries of America: refineries hungry for crude. In exchange we give electronic transfer of funds. Promises to pay from a land that consumes much but produces little. The average American adult destroys his own weight in hydrocarbons every week.

I dropped off to sleep and dreamed that I was young again and was back in El Dorado, Arkansas, for the 1968 startup of the viscous polypropylene unit. At this plant, the heat of reaction was removed by refrigeration, to maintain a reaction temperature of 40°F. At a higher reaction temperature, the viscosity of the polypropylene product was too low. But the refrigeration compressor was fully loaded in the sense that the compressor suction throttle valve (see Figure 12-1) was 100% open.

The motor amperage load for the compressor's constant-speed driver was also limiting, but only during the hot afternoons, due to the elevated cooling-water temperatures of the compressor discharge condensers (see Figure 12-2).

*Process Engineering for a Small Planet: How to Reuse, Re-Purpose, and Retrofit Existing Process Equipment,* By Norman P. Lieberman
Copyright © 2010 John Wiley & Sons, Inc.

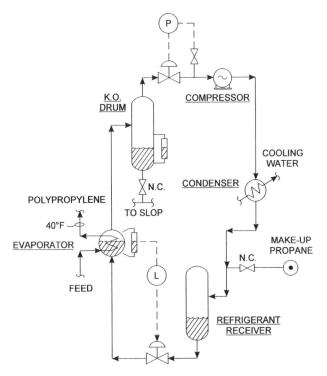

*Figure 12-1*    *Refrigeration loop.*

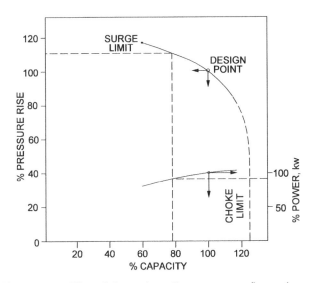

*Figure 12-2*    *Effect of change in suction pressure on flow and power.*

## EXPANSION REQUIREMENTS

The reason I often dream about this project is that it was my first design. The objective was to expand capacity by 60%. The polypropylene product was used as an additive to gasoline. It was a top cylinder lubricant used by Texaco, which purchased Amoco's entire production. Everyone at the Chicago engineering headquarters presumed that a new refrigeration compressor would be needed, as the unit was currently limited by this compressor's capacity.

It never occurred to me that I had the authority to specify a new compressor. It would cost hundreds of times my own annual salary. So my approach was to:

- Raise the suction pressure of the compressor but without raising the evaporator temperature above 40°F.
- Lower the compressor discharge pressure by improving the performance of the compressor discharge condensers.

## DISCUSSION WITH UNIT OPERATORS

The operators in El Dorado described the unit limitations to me in greater detail. The lead operator, Bill Smith, explained, "Son, our big problem is the carryover of refrigerant propane into the compressor suction knockout drum. You see, the evaporator has a tendency to puke-over. I can't understand why. We keep the level of refrigerant only half full in the gauge glass. Still, the propane refrigerant carries over into the K.O. drum. Then we got to drain the K.O. drum down to slop. But the propane refrigerant just flashes out of the slop tank to the flare. Then we lose the level in the refrigerant receiver (Figure 12-1) and have to open the propane makeup valve."

"Mr. Smith, why don't you hook up a hose and drain the refrigerant propane from the K.O. drum back into the evaporator?" I asked.

"Look, son," Bill laughed, "I guess you ain't none too smart. The pressure in the K.O. drum is lower than in the evaporator pressure. Otherwise the vapor won't flow from the evaporator to the compressor suction. Didn't they teach you in that there engineering school that water don't flow uphill?" Bill thought for a while. "But explain to me, Mr. Norm, why the propane carries up out of the evaporator when it's only half full."

"Well, Bill, that's because the density of the boiling propane inside the evaporator shell is lower than the density of the stagnant propane refrigerant in the outside level glass. You know, as the propane evaporates, it boils. And the bubbles mix up with the liquid and make the liquid less dense. But the propane in the insulated gauge glass is just flat liquid. It's kind of like when you boil water in a pot. The pot is half full, but then it carries over when it boils. If you stop the heat, the boiling water level sinks down, kind of like in the level glass."

"By cracky, you make sense," Bill exclaimed. "Maybe you are a smart engineer. It is a problem, though. When I lower the level in the evaporator to stop the carryover, the polypropylene outlet gets too hot—above 40°F. Then I have to cut back my feed. Then my compressor can suck down my pressure in the evaporator shell, which chills the evaporator back down to the 40°F. I need that 40°F evaporator temperature to make my polypropylene product viscosity on spec. If the evaporator gets too warm, my product viscosity gets too low."

"Right, Bill," I said.

"If I didn't have to worry about this carryover problem, I could make me more product. What's Texaco do with all this stuff, anyway?" Bill asked.

"Gasoline additive. Top cylinder lubricant. Mostly I think it's just a Texaco advertising gimmick," I answered. "But Bill, look here, why can't we just pump the refrigerant out of the K.O. drum and back into the evaporator instead of draining it to slop?"

"Because, Mr. Engineer, we ain't got us no pump. Maybe you should get us one," Bill suggested.

## ENGINEERING ANALYSIS

I now had a flash of brilliance! Why not install a pump on the bottom of the K.O. drum? Not only to pump back the entrained propane refrigerant but to circulate the refrigerant through the evaporator.

The evaporator is similar to a steam boiler except that we are boiling propane instead of water. Some steam boilers evaporate stagnant water and have a low heat transfer efficiency. Other steam boilers evaporate circulating water and have a large heat transfer efficiency. Heat transfer efficiency is referred to as $U$. The overall heat transfer coefficient is defined as follows:

$$Q = UA \, \Delta T \qquad\qquad (12\text{-}1)$$

where
  $Q$ = boiler or evaporator duty
  $A$ = heat transfer surface area
  $\Delta T$ = temperature difference between the heating medium and the boiling fluid

If I could increase $U$, I could increase $Q$. Also, if I could keep the evaporator tubes submerged completely in the refrigerant, the effective area for heat transfer in the evaporator would also increase. Referring to *Steam-Plant Operations* [1], I found that converting a boiler from a stagnant to a circulating mode of operation, might double the overall heat transfer coefficient. I could then transfer more heat without increasing the $\Delta T$ value in equation (12-1).

## CALCULATING REFRIGERATION COMPRESSION WORK

The problem with my plan was the capacity of the compressor. If I doubled the feed to the evaporator, the flow of vaporized propane refrigerant to the compressor would also double. But I did not have any compressor capacity to spare. What to do?

Let's consider the formula for compression work ($W$):

$$W = NR \frac{T}{A} \left[ \left( \frac{P_2}{P_1} \right)^A - 1 \right] \tag{12-2}$$

The $A$ term above is calculated as follows:

$$A = \frac{K - 1}{K}$$

Let's examine these terms. $K$ is the ratio $c_p/c_v$. For propane refrigerant, $K$ is 1.14. Therefore,

$$\frac{K - 1}{K} = \frac{1.14 - 1.00}{1.14} = 0.123$$

$R$ is the gas constant. The $K/(K - 1)$ coefficient is constant. $T$ is the suction temperature in degrees Rankine. $N$ is the number of moles of gas compressed, that is, the circulating rate of the refrigerant. Rewriting equation (12-2) yields

$$W \propto NT \left[ \left( \frac{P_2}{P_1} \right)^{0.123} - 1 \right] \tag{12-3}$$

The term $P_2/P_1$ is the compressor discharge pressure divided by the compressor suction pressure. The larger $P_1$, the compressor suction pressure, is, the less work is required by the compressor to compress the circulating refrigerant. But that is not the main benefit of raising $P_1$. This compressor was not normally limited by work or excessive motor amps anyway. Except on hot days, the motor amp load was not excessive. The normal limit was that the compressor suction throttle valve was wide open. The problem was that the compressor was limited not by equation (12-2) but by the amount in pounds that would flow into the compressor suction. I call this a *suction volume limitation.*

## HORSEPOWER LIMITED VS. SUCTION VOLUME LIMITED

I dreamed on about doing engineering calculations. In my dream I discriminated between two compressor limitations:

- *Horsepower limitation.* For a constant-speed motor-driven centrifugal compressor, this means that the motor driving the compressor is at the maximum amperage load, beyond which it will trip off. If the driver was a steam turbine, it would mean that the governor speed control valve would run wide open.
- *Suction volume limitation.* For a constant-speed motor-driven centrifugal compressor, this means that no more gas will flow into the compressor at the speed at which it is running. If the driver was a steam turbine, the governor would be controlling the compressor to run at its maximum rated speed.

Referring to Figure 12-1, if the compressor was over-amping, the suction pressure control valve would have to be throttled to reduce the refrigerant flow and, consequently, the motor amps. If the compressor was limited by suction volume, the pressure control valve on the suction would be 100% open. Let me remind the reader that this summary is only for a constant-speed compressor.

## EFFECT OF INCREASING SUCTION PRESSURE

Let's assume that $P_2$, the compressor discharge pressure in equation (12-3) is constant. Now let's assume further that I have somehow increased the compressor suction pressure, $P_1$. Let's assume that I have somehow doubled the suction pressure. Thus, the compression ratio $(P_2/P_1)$ has been cut in half. But if the volume of gas flow remains the same, the number of moles of gas compressed, $N$, will double; that is, $N$ in equation (12-3) will double.

If I cut the compression ratio in half but double $N$, you can calculate that the effect on the change in work required to drive the compressor is small. Actually, I am oversimplifying a complex problem. Centrifugal compressors run on curves, as shown in Figure 12.2 [2]. As you may calculate from the dashed straight lines I have drawn on the compressor curve:

- I have raised the compression ratio by 10% over the design point (to 110%).
- As a result, the capacity for gas flow has dropped by 22% below design (to 78%).
- Also, the percent of the power has dropped by 10% below design (to 90%).

Note that there is a trade-off between decreasing the compression ratio and increasing flow and decreasing compressor work—but the trade-off is not linear. I realize that this is complex, but if you want to understand my point, you will have to play with equation (12-3) and Figure 12-2 by inserting several different compression ratios.

In the 1960s, I did not understand anything about compression curves or how to calculate compression work. But, I could see that raising the compressor's suction pressure had to give me more refrigerant circulation capacity. But how could I raise the compressor suction pressure?

That was easy. I would just allow the evaporator temperature to increase. Not on the polypropylene polymerization side of the evaporator (i.e., the tube side) but on the refrigerant side of the evaporator (i.e., the shell side).

But how could I avoid exceeding the 40°F maximum on the process feed or polymerization side of the evaporator? The answer lies in equation (12-1). By circulating the propane refrigerant through the evaporator, as shown in Figure 12-3, I would increase both $A$ and $U$ in equation (12-1). The heat transfer surface area ($A$) would go up because all the tubes would always be completely covered and submerged in liquid refrigerant. Of greater import, the $U$ value for circulation of the boiler would perhaps be double the $U$ value for the current stagnant propane refrigerant in the evaporator.

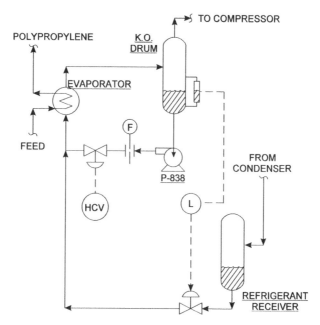

**Figure 12-3**   *Converting a once-through evaporator to a circulating evaporator to increase heat transfer efficiency.*

I could then increase the pressure and temperature of the refrigerant in the evaporator. It's true that the $\Delta T$ term in equation (12-1) would be diminished, but I would more than offset this reduced $\Delta T$, with a bigger $U$ and a bigger $A$.

The only problems I had were that I had no idea how to do the calculations. Also, sometimes the compressor was not limited by suction volume but by the maximum motor amperage loading.

## REDUCING REFRIGERANT CONDENSER FOULING

To overcome the motor overload case, I added several facilities to my condenser to permit three additional operations:

- *Air blowing.* High-pressure air is blown through the tubes for 10 minutes every day. Typically, this increases water flow by 10% and retards fouling rates by blowing cooling-water sludge out of the tubes.
- *Back-flushing.* Water flows backward through the tubes to remove obstructions stuck onto the tube sheets. The water supply valve is closed. A connection on the water supply inlet, or channel head, is opened to drain cooling water to the sewer. For this to be effective, the drain connection must be at least half the diameter of the water supply line. Back-flushing is typically done once a week for about 10 to 20 minutes, or until the back-flushed water runs clear.

- *Acid cleaning.* A dilute acid (such as inhibited hydrochloric acid) is injected through a 1-in. connection in the cooling-water supply line. Caustic is injected into the water discharge to neutralize the effluent pH. Carbonate deposits are removed completely in a matter of minutes. This is very effective in restoring cooling-water flow and heat transfer coefficient, but corrosive to the carbon steel cooling-water return header. Typically, this is done only once each summer.

Reducing the compressor discharge pressure would reduce the compressor suction pressure, but without changing the compression ratio ($P_2/P_1$) or the number of moles of refrigerant ($N$) compressed. This would reduce the evaporator temperature with no increase in the motor amperage load or loss in propylene polymerization capacity.

My bigger problem was that I had no idea how to predict the effect of these changes. I would need to use the compressor's head vs. flow curve (see Figure 12-2). But, I did not know that such a chart existed. Equation (12-2) was also unknown to me. In any case, could I really quantify the effects on the evaporator area or heat transfer coefficient by installing the refrigerant circulating pump? Not really. Also, I didn't know how to quantify the effect of reduced fouling in the condenser by air blowing, back-flushing, and acid cleaning.

So I decided to work backward. The required increase in capacity was 60%. I assumed that the compressor absolute suction pressure would be 60% higher because the refrigerant vapor leaving the evaporator would be 15°F warmer. The warmer refrigerant would reduce $\Delta T$ in equation (12-1) but that would be offset, as I explained in my report, because of the new pump.

I still have my old design report; it is complete process engineering nonsense. But no one at Amoco read it, so my design was approved. The circulating pump was installed and the condenser was acid cleaned. When the unit was re-streamed, capacity increased by exactly 60%. Doesn't this prove that I was inspired by the Creator of the Universe?

Suddenly a blast from a passing tugboat's horn awoke me. The mighty river still flowed at my feet. Hundreds of miles upstream, the Mississippi was joined by the Arkansas River, which flowed past the memories of my youth. And in case you're thinking I made this story up—not the refrigerant part, but the youth part—Figure 12-4 is a photo of me in El Dorado during the startup. The meter on my right is the circulating refrigerant flowmeter shown in Figure 12-3. I'm sitting on P-838, the new refrigerant circulating pump.

## ADJUSTING REFRIGERANT COMPOSITION

I suppose some readers are thinking that the entire project was Bill Smith's idea. Not true. The part of the project that was my contribution was adjusting the volatility of the refrigerant. You see, propylene has a higher vapor pressure than propane. By spiking the circulating propane refrigerant with the more volatile propylene, the compressor suction pressure increased. Therefore, even though the compressor suction

**Figure 12-4** *El Dorado, Arkansas viscous polypropylene unit startup, 1969. The HCV Figure 12-3 is at my right. I've just opened it 100%.*

throttle valve was spun wide open, we could still increase the pounds of refrigerant circulated.

The compressor discharge pressure also increased, because the volatile propylene is more difficult to condense than the propane alone. We overcame this problem by acid-cleaning the condenser more frequently, to reduce the compressor discharge pressure. Also, we reduced the cooling-water supply temperature a few degrees by clearing the plugged water distribution holes in the top deck of the cooling tower.

## SUMMARY

In Chapters 10 and 12, I have told similar stories. In this chapter, I purchased a small pump and improved maintenance of the condenser. This avoided the purchase of a 1000-hp compressor to replace the existing 600-hp compressor. Not only that, but the extra energy required to run the larger machine was also saved.

In Chapter 10, I installed a piece of pipe inside the bottom of the crude tower stripping section to avoid construction of a new diesel oil recovery tower. Not only was a huge construction project negated, but the energy to run the new tower was also saved.

The successful outcomes from such projects very early in my career have molded my approach to engineering in general. After all, we live on a little speck of dust, floating in a remote corner of our universe. But it's our only home. So we had better use what we have, avoid building new stuff that we don't really need, and avoid the burning of unnecessary fossil fuels.

## EFFECTS OF NONCONDENSIBLES IN CIRCULATING REFRIGERANT

In refinery applications, the circulating refrigerant is not normally Freon or ammonia, but the refinery process stream itself. It's sort of a form of autorefrigeration. But still, the refrigerant must be compressed, condensed, and then recirculated back into the process loop. An example of this is the sulfuric acid alkylation ("alky") unit used in many refineries.

Himmler Oil operates a large refinery in New Jersey. For many years, their refrigeration loop on their sulfuric acid alky unit was limited by high compressor discharge pressure. The high discharge pressure resulted from a warmer cooling-water temperature during the summer (see Figure 12-2). To avoid this limitation, Himmler Oil employed a portable rented $NH_3$ refrigeration unit to provide refrigerated water to a final trim condenser during summertime operations. Using chilled, refrigerated water to help condense a refrigerant is a truly inefficient may to provide extra refrigeration capacity. As the driver for the compressor on the chilled water system was an electric motor, the inherent inefficiency was multiplied by 3. That is, it takes 3 to 4 Btu of energy to generate 1 Btu of shaft horsepower from an electric motor.

This sort of chilled water system is also inherently inefficient because:

- The circulated $NH_3$ refrigerant must be compressed and then condensed.
- The evaporating $NH_3$ must chill the water, which will then be circulated by a separate pump.
- The chilled water must then condense the residual, volatile hydrocarbon vapors, in the alky refrigeration loop.

If the residual hydrocarbon vapors are not condensed, the pressure in the refrigerant receiver shown in Figure 12-5 would increase. This would push up the liquid level in the condenser. The saturated refrigerant would become progressively subcooled.

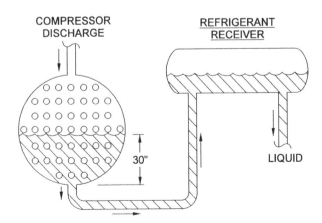

*Figure 12-5*   *Condenser area halved due to condensate backup.*

That is, the refrigerant liquid would be subcooled enough that when it entered the refrigerant receiver, it would be cold enough to absorb all of the lighter components in the receiver. The fact that the refrigerant receiver at the Himmler refinery was located above the condenser just made the problem of condensate backup worse.

The tubes that were submerged in liquid were not condensing any vapor. Therefore, as the liquid backed-up in the condenser, the rate of condensation dropped. This raised the compressor discharge pressure, which reduced the refrigeration circulation rate (see Figure 12-2) and hence the alky feed capacity. My client had a thermographic infrared photo taken of the condenser. Over 50% of its surface area was submerged, when the chilled-water flow to the secondary condenser was stopped (see Figure 12-5).

A sample of the gas phase in the refrigerant receiver showed it to be mainly methane, ethane, and propane. These moles were probably introduced in very minor concentrations with the butane and butylene feeds and became trapped in the refrigerant. We then opened the vent on top of the refrigerant receiver. The amount of gas vented was too small to register any loss of the recirculated liquid refrigerant as indicated on the flow meter. The vented gas was reprocessed in the adjacent saturated hydrocarbon gas recovery plant. The effect observed on the alky unit is summarized in Figure 12-6:

- The pressure in the refrigerant receiver began to drop.
- The subcooled liquid level in the condenser, as observed in the thermographic infrared photo image, dropped from 30 in. to about 12 in.
- The outlet temperature of the condenser increased, due to the reduction in refrigerant subcooling.

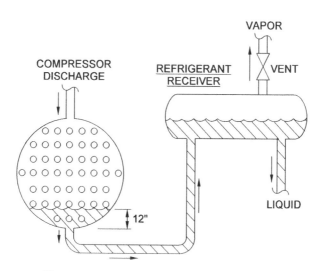

**Figure 12-6**  *Venting reduces condensate backup.*

- The compressor discharge pressure also declined.
- The refrigeration circulation rate increased, due to the reduced compressor discharge pressure, as shown in Figure 12-2.

## THE WRAPUP MEETING

I don't know why I go to meetings. I always wind up getting angry and use language that I later regret. Steve Hill, the tech manager at Himmler, forced me to go to this stressful meeting in New Orleans with the engineering contractor firm MERS. The contractor's calculations indicated that a new compressor rotor was required for the alky unit refrigeration machine. The new rotor would be able to produce 25% more polytropic head than the existing compressor's rotor. With the new rotor, the rented chilled water system could be eliminated, the MERS rotating equipment engineer explained.

Why didn't the nationwide MERS Corporation realize that the problem was condensate backup in the refrigerant condenser? Why did they confidently advise Himmler Oil to embark on another senseless project which would chew up the resources of our little planet? Well, there are two possibilities:

- The MERS engineers never leave their air-conditioned offices in the French Quarter to investigate the true situation in dreary New Jersey, or
- The MERS Corporation has been taken over by a malevolent entity wishing to destroy our planet.

I'm sure that my readers are wondering, "How come so much evil engineering is concentrated in New Orleans?" My response is simple. Evil in New Orleans is not limited to process engineering. It's just that we live 10 ft below sea level. Each day could be our last. Sodom and Gomorrah had the same problem.

## REFERENCES

1. Woodruff, E. B., and Lammers, H. B. *Steam-Plant Operations*, 4th ed., McGraw-Hill, New York, 1977.
2. Bloch, H. *A Practical Guide to Compressor Technology*, 2nd ed., Wiley, Hoboken, NJ, 2006.

# Chapter 13

# Oversizing Equipment Pitfalls

I am now a tax-exempt entity. PET (Process Engineering Temple) of New Orleans has received a certificate from the IRS recognizing us as a not-for-profit establishment. Your contributions to PET are tax deductible. As an established religion, we hold regular services. My best friend, Mark, flew in from Dallas to attend. After my sermon, Mark rushed up to me. "Norm," Mark said, "Everything you said was totally wrong. Especially about us being friends."

## SCRUBBING H$_2$S FROM HYDROTREATER RECYCLE HYDROGEN

I had spent many weeks working in the Carib refinery, mostly doing retrofit designs and field troubleshooting. Mark was the tech service manager. My sermon harked back to those sunny days on that happy island.

"Fellow devotees of technology. Thank you for coming today. My specific sermon for this Sabbath is design of hydrogen gas recycle for H$_2$S amine absorbers. The more general subject is following standard design practices that make no sense."

"Norm," Mark interrupted, "I thought you invited us here for lunch. When are we going to eat? Where's the beer?"

"Patience, brother Mark," I responded. "Dear friends, I recall the day I was sitting quietly in my office in Carib when Kevin Rains, the maintenance supervisor, burst in."

*Process Engineering for a Small Planet: How to Reuse, Re-Purpose, and Retrofit Existing Process Equipment*, By Norman P. Lieberman
Copyright © 2010 John Wiley & Sons, Inc.

"Lieberman, where's that other idiot?" asked Mr. Rains.

"Kevin, are you perhaps referring to my dear friend and colleague, Mark?" I inquired.

"Yeah! That idiot. We just opened up T-308, the amine scrubber."

"How does it look? Any fouling on the trays?"

"Yeah! A bunch of black gritty material is on all the trays that ain't completely eaten out. What's that black stuff anyway, Lieberman? Coke?"

"No. It's iron sulfide [i.e., $Fe(HS)_2$]. Not coke."

"You mean pyrophoric iron? The stuff that autoignites when it dries out and is exposed to air? Is that right, Lieberman?"

"Just so. When it burns, it evolves into sulfur dioxide ($SO_2$), which is fatal in concentrations over 1000 ppm."

"We'll wash it out right away."

"I don't think so. Pyrophoric iron is insoluble in water and hydrocarbons. You need to clean the tower chemically with a weak acid solution to make it safe to enter. What do you want with Mark anyway?"

"Okay, I'll chemically clean the tower with a "zyme" solution. But did you know that 16 of the 20 trays are eat-up?"

"The word is 'corroded,' not 'eat-up.' But yes, that was reported six months ago during the emergency turnaround in May. That's why Mark ordered 20 new replacement trays of upgraded design and enhanced capacity. Further, people who do not use proper syntax should not refer to persons that do so as 'idiots.' "

"Look, Norm, here's my problem," Kevin said more quietly. "The new trays are two-pass trays. The existing trays are single-pass. That's why I'm so mad."

This was bad news. "I'm sorry, Kevin. What can I do to help? Didn't you know about this before?"

"Nope. Mark didn't tell me."

Modifying a tray support ring from single-pass trays to two-pass trays is a complex business (see Figure 13-1). More serious would be welding onto the vessel wall the 80 new vertical downcomer bolting bars in the small confines of the 5-ft-diameter tower. This would be an awful, time-consuming task.

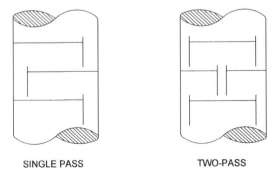

SINGLE PASS                    TWO-PASS

**Figure 13-1** *Changing the number of tray passes is mechanically complex and is best avoided when possible.*

"But look, Norm," Mr. Rains said in a friendlier tone, "I've been speaking to Mr. German in Operations. He said that at 12 million scf/day of recycle gas, the tower flooded."

"Right," I agreed. "The trays are flooding because they are fouled with iron sulfide deposits. But Mark's new trays are designed for 20 million scf/day. That's why they are two-pass trays. Also, they are a higher-alloy steel—410 chrome steel, instead of carbon steel."

"Right. That's good," Kevin hesitated. "But look, Lieberman, Mr. German said at 10 million scf/day of recycle gas, the tower didn't flood, and scrubbing was okay, too."

"So what's the point?" I asked.

"My point is that if scrubbing was okay, and I have only four trays in the tower now because the other 16 trays are eaten . . ., excuse me, corroded, then why do I need to install all 20 of the new trays? Why not just replace these four old intact trays with the new alloy two-pass trays?"

"I told you. Mark ordered the new trays for the extra capacity to handle the extra gas flow." But I already knew my answer was nonsense.

"Look, Lieberman. I'm not a process guy, but aren't you getting confused between gas capacity and $H_2S$ removal efficiency, which is a function of the number of contacting stages?"

"Kevin, what do you want anyway?" I could feel little bumps forming on the back of my hands.

"Give me written approval to install just four new trays and the seal pan. That's it."

"Mark will be back Monday," I said.

"I need the letter today. I sure can't wait till Monday."

"Look. This is a standard Easton Oil design. They always design amine absorbers with 20 trays (see Figure 13-2). They have a thousand of these towers, all with 20 trays. They have been designing these towers with 20 trays since World War II. Everybody uses 20 trays in amine absorbers. It's standard."

"That don't make it right, Mr. Lieberman."

"Look, Kevin, could you come back later?" I asked. "Let me think about it for awhile."

"Sure. I'll be back at one. I'm just on my way to lunch."

## OPTIMIZING H₂S ABSORPTION TRAYS

I knew the drill. If I didn't give Mr. Rains the answer he wanted, he would go right up to see Mr. Haynes, the general director. Mr. Haynes would trot down the stairs to see me within 2 minutes. Then I would still have to decide about the correct number of absorption trays in T-308. And the answer could not be because that was an Easton Oil best practices design. Mr. Haynes would never accept such an answer.

The gas oil feed to the desulferizer reactor contained 3 wt% sulfur. As a result, the untreated hydrogen recycle gas had an $H_2S$ content of 8 mol% sulfur, or 80,000 ppm

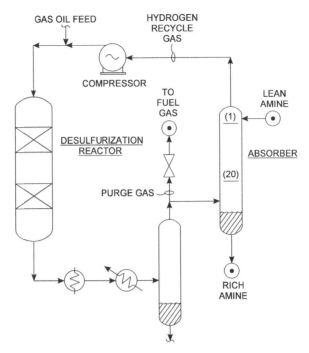

**Figure 13-2** *Hydro-desulfurizer recycle gas loop.*

$H_2S$ by volume. When it was properly stripped and circulated at an adequate rate, the lean amine could reduce the $H_2S$ content of the recycle gas from 80,000 ppm down to 30 ppm in an amine absorber with 20 trays. By calculation and field experience, I knew that reducing the number of absorption trays from 20 to five would only increase the $H_2S$ content of the recycle gas from 30 ppm to 100 ppm. Indeed, I had published the results of such a plant test in a book I authored [1]. And Mr. Haynes had read this book, because I had foolishly given him a free copy.

The problem was not really with my book. The real problem was Mr. German. He had told Mr. Rains that as long as the $H_2S$ content of the hydrogen recycle gas was not more than 1000 ppm, he didn't care. Just to make sure, I phoned Mr. German in the control center.

"Yes, Mr. Norman, it is so. I told our good friend Kevin that the $H_2S$ amine absorber must remove most of the $H_2S$ in the recycle hydrogen gas. Otherwise, its concentration will build up in the recycle gas loop. The $H_2S$ content of the recycle gas, whether it's 100 ppm or 1000 ppm doesn't matter. As soon as hydrogen gas mixes with the high-sulfur gas oil feed and touches the first foot of catalyst in the reactor, its $H_2S$ content jumps to thousands of ppm. So it doesn't matter if the hydrogen is 10 ppm, 100 ppm, or 1000 ppm. Even at 1000 ppm, the absorber will still be extracting 99% of the $H_2S$ produced in the reactor, which is quite enough. That's what I told Kevin. And how are you today, my friend?"

"I'm good. But I guess from what you've said, we don't need to install all 20 trays. Maybe five or 10 trays will be enough. I mean, you were running okay, as far as absorption efficiency goes, with just four badly corroded trays for the past few years. Maybe 10 trays are enough?"

"No! No! No! my good friend Norman. We must return all 20 trays to the absorber during the turnaround. No, my good friend, all 20 trays are certainly needed."

"But why?"

"Because, Mr. Norman, that's an Easton Oil standard design practice."

Now what?! Why did Easton Oil have to use 20 trays in the first place? Maybe one might need all these trays to scrub refinery fuel gas down to 20 ppm of $H_2S$. But in a gas oil desulfurizer recycle loop scrubber, why didn't Easton reduce the number of trays in a logical fashion? It just wasn't the waste of the tower height and the trays. The extra trays imposed an unnecessary $\Delta P$ on the recycle gas, which wasted the recycle gas compressor horsepower. Even with just four intact trays, I would still reduce the $H_2S$ in the recycle gas to about 100 ppm. What to do?

I decided to compromise. Kevin would be back from Wendy's in 15 minutes, so I wrote this letter:


To: Mr. Kevin Rains
     Maintenance Sup't.
     Hydroprocessing Division

This letter authorizes you to replace the existing 20 single-pass trays in T-308, with eight new two-pass trays. New downcomer bolting bars and seal pans are required. Modify existing tray rings as per vendor drawings. 410 S.S. required.

Norman P. Lieberman
Consultant Engineer


## RESULT OF REDUCED ABSORBER TRAYS

When Mark returned to work Monday, he was furious with me for deviating from his design for the $H_2S$ absorber. Mr. Rains was angry, too. He claimed that I forced him to do twice the necessary work. As proof that only four trays were needed, he taped the lab analysis on my office door showing 50 ppm of $H_2S$ in the recycle gas with a red note: "Is Eight Enough?"

Mr. German was also upset. He wanted his 20 trays because that was an Easton Oil best practice design. And Mr. Haynes was insulted that I had not consulted with him on such a critical process equipment decision.

So I went down to the beach, removed my shoes, sat in the sun, and had a beer.

"Fellow devotees of technology, that concludes my sermon for today. Let us always remember the words of Kevin Rains, 'Just because things have been done a certain way for a long time, does not make it right.' Amen."

(*Safety Note*: The author recommends use of a Permanna cleaning program. This is a two-step procedure. The first step is circulation of a solution that removes residual heavier hydrocarbons from a tower's trays. The second step circulates a solution that chemically removes the iron sulfide deposits. The evolution of sulfur dioxide fumes inside a vessel is a serious safety problem, as the author has learned from a bad personal experience at the Texas City sulfur plant in 1980. $SO_2$ is just as deadly as $H_2S$ in a confined space!)

## ABSORBER OVERDESIGN

In general, many amine absorbers have too many trays. As I noted in Chapter 5, too many trays may lead to excessive absorption of $CO_2$, which then wastes amine circulation and sulfur plant capacity. Also, too many absorption trays results in a temperature bulge in the middle of the absorber. The temperature at the top of the absorber is suppressed by the cold lean amine. The temperature at the bottom of the absorber is suppressed by the cooler inlet gas. Hence, the temperature toward the middle of the absorber tends to be higher than the top or bottom temperatures. This temperature bulge is magnified by having more absorption trays, in that the center of the tower is farther away from the top and bottom of the tower. This temperature bulge can be quite extreme. I have observed that the midpoint temperature of an absorber can be 60°F (i.e., at almost 200°F) hotter than either the top or bottom trays. Corrosion rates in amine systems might be 10 times higher at such elevated temperatures. Also, the midsection of vessels are not areas expected to have the highest corrosion rates, and safety inspections may often be omitted in such areas.

## CONSEQUENCES OF OVERDESIGN

In general, amine hydrogen sulfide absorber overdesign is just an example of the problem of overdesigning process equipment. Not only is it a wasteful practice, but other problems are created as well. My favorite example of the wastefulness of overdesign is distillation tower trays. Not only is the vessel diameter increased for no purpose, but the oversized trays suffer from low vapor velocity. This causes the $\Delta P$ value of the vapor flowing through the tray deck orifices or caps to be too low. Consequently, the tray decks start to weep, or leak, or dump. Unless the trays are dead-level, vapor–liquid channeling will result [2]. This will cause the tray efficiency to be adversely affected. The lower tray efficiency will require more reflux to make the required split. More top reflux will require more steam flow to the reboiler. This wastes energy and promotes the evolution of $CO_2$ into the atmosphere. Oversizing equipment past the best efficiency point of the tray, the pump, or the compressor hurts the energy efficiency of the process unit.

## OVERSIZING VAPOR–LIQUID SEPARATORS WITH DEMISTERS

In Chapter 20 I discuss the importance of a demister in suppressing an acid mist from a sulfuric acid plant. However, I first encountered the critical importance of demisters in two other services:

- As an final condenser in a sulfur recovery plant to remove a sulfur mist.
- In an amine $H_2S$ fuel gas absorber to remove an amine mist.

Ordinarily, one would think that a larger diameter knockout drum (Figure 13-3) would do a better job of allowing entrainment to settleout of a gas stream, as discussed in Chapter 15. According to Stokes' law, droplet-size entrainment is proportional to vertical gas velocity squared. That is, if the cross-sectional area of a vessel is reduced by half, liquid droplets four times as massive could be entrained in the flowing vapor. Experience proves that this is directionally correct. However, to remove the very fine liquid droplets in a mist or fog efficiently, a mechanical device to coalesce the droplets is required. This device is called a *demister*.

Without efficient removal of the entrained liquid sulfur mist in the final condenser, conversion of the sulfur plant would be adversely affected by several tenths of a

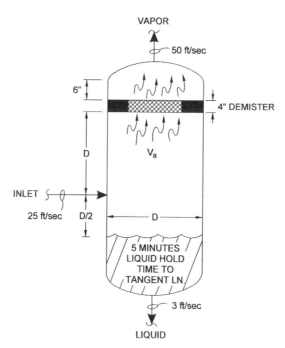

***Figure 13-3*** *Oversizing demisters promotes entrainment losses.*

weight percent. If a sulfur recovery plant lacks a tail gas recovery section, this is a serious environmental debit.

Without efficient removal of an entrained liquid amine mist in the off-gas from an $H_2S$ scrubber, amine will contaminate the refinery fuel gas. Burner tips in the heaters will plug and result in wasted fuel, as more air is needed to offset the loss of air–fuel mixing efficiency in the plugged burner tips. Amine carryover will settle out in the fuel gas K.O. drums. Lost amine will overload the refinery's effluent treatment plant with organic nitrogen compounds.

## HOW DEMISTERS WORK

A typical demister is 4 to 6 in. thick. High-velocity droplets impinge on the demister's fine fibers with substantial momentum. If the droplet's momentum is too low, the finer droplets will not coalesce into larger drops of liquids. Then the low-velocity entrained droplets drift through the demister without coalescing. Hence, low vapor velocity degrades the demister's liquid separation effectiveness.

On the other hand, excessively high vapor velocities will blow the droplets of liquid right through the demister's fine fibers. Thus, demisters will work optimumly within a relatively narrow range of velocity:

$$V_a = K \left( \frac{D_L}{D_V} \right)^{1/2} \tag{13-1}$$

where
   $V_a$ = vertical vapor velocity, ft/sec
   $D_L$ = liquid density
   $D_V$ = vapor density
   $K$ = 0.22 to 0.28 ft/sec

## EFFECT OF OVERSIZING DEMISTERS

The cross-sectional area for both the sulfur plant final condenser and the amine fuel gas scrubber are typically double or triple that required to produce the optimum demister gas velocity. Thus, it is often necessary to install a partial blanking plate to reduce the cross-sectional area of the vessel (Figure 13-3) to obtain an adequate vertical gas velocity for efficient demister operation. The size of this blanking plate is calculated from $V_a$, as defined in equation (13-1).

Alternatively, I recall a Catacarb $CO_2$ absorber in a hydrogen plant. The absorber suffered from massive continual losses of the Catacarb absorbent solution. The $K$ factor, as defined above, was in excess of 0.5 ft/sec. I corrected this entrainment loss by doubling the cross-sectional area of the existing demister.

## DEMISTER FAILURE

Unfortunately, I suffer from the rare disease F.O.D. (Fear of Demisters). I contracted this disease in 1974 when a demister failed in a centrifugal compressor interstage knock-out drum on a unit I was supervising in Texas City. The demister became lodged on the compressor's second-stage rotor and led to a shutdown. Two errors had been made in the design of this demister:

- The use of demisters should be viewed as a calculated risk, as they are subject to plugging and to mechanical failure. In a centrifugal compressor, carryover of mist is beneficial for the downstream rotor. It keeps the rotor wet and clean. On the other hand, carryover of amine from an $H_2S$ fuel gas absorber should best be minimized with a demister.
- The material of construction of the demister must not be subject to corrosion. The products of corrosion will foul the demister and lead to an increased $\Delta P$. A pressure drop of 1 or 2 psi is sufficient to rip a demister away from its wall supports.

The other reason I suffer from F.O.D. is that a partially plugged or coked demister is far worse than no demister. I recall the visbreaker fractionator at a plant in Louisiana. Black gas oil was being produced due to entrainment of residual components. When I inspected the demister, it was 100% solid coke. A small corner of the demister had pulled away from the vessel wall. Through this tiny opening, all of the vapor flowed at a high localized velocity. The refinery removed the demister. But contrary to their practice for the past 20 years, I asked them not to replace it. Without the demister, gas oil color quality was sustained for an extended run length. Caution! To avoid contracting F.O.D., use demisters only when they are really needed. Excessive demister use may be hazardous to your health.

I have shown in Figure 13-3 how I size knockout drums for minimum entrainment. Again, note that 90% of the time I just size my vessels for gravity settling, without the use of a demister. See Chapter 15 for details on sizing knockout drums for a variety of services. Nozzle velocities shown are typical but are governed by permitted pressure drops.

## REFERENCES

1. Lieberman, N. P. *Troubleshooting Process Operations*, 4th ed., PennWell Publications, Tulsa, OK, 2009, pp. 429–434.
2. Kister, H. Z. *Distillation Operation*, McGraw-Hill, New York, 1989.

# Chapter 14

# Optimizing Use of Steam Pressure to Minimize Consumption of Energy

All types of heat, like all types of people, are not created equal. For example, I have been waiting for 46 years to play for the NBA. I practice every day. But so far, my name has never come up in the NBA draft.

Energy is like that. At the top of the heap we have electricity. That's like the Kobe Bryant of energy. At the bottom we have warm water from a process cooler. That's like the Norm Lieberman of energy. To make a Btu worth of electrical energy takes 3 to 4 Btu of energy from burning hydrocarbons. This is done in a coal-fired power plant by raising high-pressure steam and then expanding the steam through a turbine, which drives a generator. The steam cools. Also about 10 to 20% of the steam condenses as it passes through the turbine. However, the majority of the steam condenses, and its heat content is lost in the downstream surface condenser. That's why we pay $30 per million Btu/hr (i.e., 10 cents/kWh) for electrical energy that was generated from coal that may have cost $4 per million Btu/hr, or natural gas that may have cost $8 per million Btu/hr.

You can take electricity and convert its heat content into whatever temperature you desire. But the heat in 120°F cooling water is only available at 120°F. Electrical energy can be used for every sort of energy requirement. Warm water can only be used to heat greenhouses in Lithuania. It's like me and Kobe. He can make shots from anywhere. I can sink the basketball from only one spot in my driveway.

*Process Engineering for a Small Planet: How to Reuse, Re-Purpose, and Retrofit Existing Process Equipment,* By Norman P. Lieberman
Copyright © 2010 John Wiley & Sons, Inc.

We have lots of low-level-temperature heat available in process plants. But as the temperature of heat is reduced, we have to become smarter in how we use it. I will summarize this concept in the form of the second law of thermodynamics:

$$W = \Delta H(T_2 - T_1)$$

where
   $W$ = useful energy in the form of work or heat used in a process
   $\Delta H$ = available heat or energy
   $T_1$ = temperature at which I can use the $\Delta H$ heat
   $T_2$ = temperature at which the $\Delta H$ heat is available to me

The best way to think about the second law of thermodynamics is that $T_2 - T_1$ is like a rating factor for available heat. A large $T_2 - T_1$ value means that the available heat is very valuable. A zero value for $T_2 - T_1$ means that the available heat is worthless—like my potential value to the NBA's New Orleans Hornets.

## PRESERVING THE POTENTIAL OF STEAM TO DO WORK

So how are we doing in combating carbon dioxide in the atmosphere? We have all heard about wind and solar power, ethanol, biodiesel, fuel cells, algae oil extraction, and hybrid vehicles. But what is the actual result?

I just checked on the Internet. No progress. The carbon dioxide concentration has been rising at a steady cumulative rate of 0.52 mol% since 1970. This means that your daughter born this year will have to contend with 590 ppm $CO_2$ in the atmosphere vs. the 280 ppm that prevailed for tens of thousands of years before the English invented the steam engine in the eighteenth century to pump water out of tin mines. Certainly the human race will survive the 590 ppm of carbon dioxide in the atmosphere. But equally certain, it will not be in our current numbers or our accustomed manner of existence. If you believe that the problem is long term in nature and will not really affect our planet in your lifetime, let me summarize one observation made by satellite. In the last 30 years, the extent of the Arctic Sea ice has dropped by over 30% [1]. If you had a 30% change in a process variable on a reactor, would you not be somewhat alarmed? Especially if the change occurred on your shift but not as a result of any intentional operating parameter change. We in the process industry need to start doing our job differently. So I'll give an example of some new technology that you and I can use today.

## POWER RECOVERY FROM STEAM TO A REBOILER

Figure 14-1 illustrates how we can preserve the value of steam being used in a distillation tower reboiler. The process design engineer has selected to use the

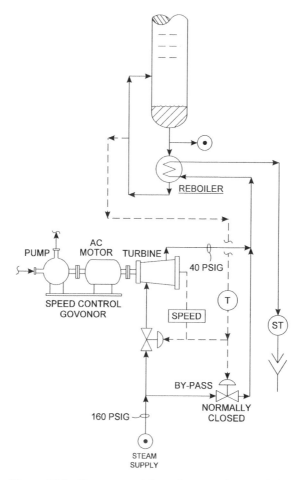

*Figure 14-1*  *Recovery work from steam supply to a reboiler.*

150-psig steam to reboil the tower, even though the required steam pressure to the reboiler is only 40 psig. Her reasons were:

1. Allowance for the $\Delta P$ value of the steam across the steam supply (TRC) control valve.
2. An alternative operation, used only 10% of the year, requires a higher tower bottoms temperature and thus greater steam pressure to the reboiler.
3. The tower must be run at 200 psig during the hottest day in August to condense the overhead product, but only at 160 psig the rest of the year, when there is colder ambient air to the fin-fan coolers. Lower tower pressure means a lower tower bottoms temperature.

  4. The refinery has only three steam supply pressures available:
     - 450 psig superheated to 600°F
     - 160 psig superheated to 400°F
     - 50 psig saturated at 300°F

Because of the first three factors, the process design engineer has correctly selected to use the 160-psig steam supply. Ordinarily, then, the steam pressure would have been reduced from 160 psig to 40 psig through a control valve. This would not reduce the heat content of the steam. But it would reduce the temperature at which the heat is available.

As shown in Figure 14-1, instead of a control valve, the steam is passing through a topping steam turbine. Work and heat is extracted from the steam as it expands from 160 psig and 400°F to 40-psig steam. But how much heat is converted to work? You must now reference your Mollier diagram in the back of your *Steam Tables*. The relevant portion of the Mollier diagram is reproduced in Figure 14-2, and an SI-unit version is shown in Figure 14-3. Proceed as follows:

*Step 1.* Plot the point 175 psia and 400°F on the chart. Enthalpy is 1218 Btu per pound of steam (i.e., vertical axis).

*Step 2.* Drop straight down the vertical constant-entropy line to the 55-psia steam line. Enthalpy is 1120 Btu per pound of steam.

This means that you have converted $1218 - 1120 = 98$ Btu per pound of steam to work. What has happened to this work? From Figure 14-1 it shows that energy in the form of shaft work is helping the turbine-spin the motor. This reduces the horsepower demand of the motor as follows:

- Assume that the steam flow to the reboiler is 10,000 lb/hr.
- Heat converted to work is then $(10,000)(98) = 980,000$ Btu/hr.
- To obtain horsepower: $980,000 \div 2457$ equals 400 hp.

But suppose that the power needed to drive the pump is only 180 hp. What will happen to the other 220 hp? Will it go back to heat? Absolutely not! It will be converted to electricity: electric power that is exported into the electrical grid. A three-phase alternating-current motor acts like a generator. If it is driven with more energy or torque than it needs to drive the pump shaft, the extra energy is converted into electricity.

But what good does this scheme do? After all, I'm just converting 98 Btu/lb of steam from heat into work. After all, I'm just stealing 98 Btu/lb worth of heat from the steam from the reboiler to drive the pump and generate electricity. Quite true. But all types of heat are not created equal, just like me and Kobe Bryant.

By upgrading this 98 Btu/lb of heat to torque that drives my pump, and to electrical power that I export, I have made this 98 Btu/lb three times more valuable. Recall that it takes 3 to 4 Btu of heat from burning hydrocarbons to make a single Btu of electric

ENTROPY

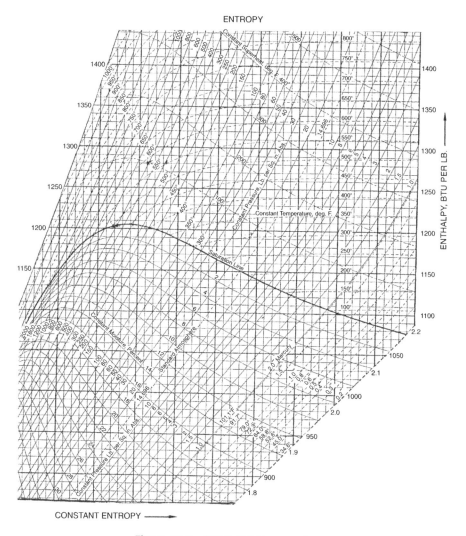

CONSTANT ENTROPY ——▶

*Figure 14-2*    *Steam enthalpy vs. entropy.*

power. In effect, I have saved $2 \times 98 = 196$ Btu/lb of steam consumed in my reboiler. The "2" factor [2] is just the difference in the energy value between electric heat and low-pressure steam.

The amount of heat needed to generate 1 lb of 160 psig and 400°F superheated steam from 180°F boiler feedwater, is about 1100 Btu/lb of steam. Thus, the facilities shown in Figure 14-1 will save:

$$196 \div 1100 = 18\% \text{ of the equivalent reboiler steam supply}$$

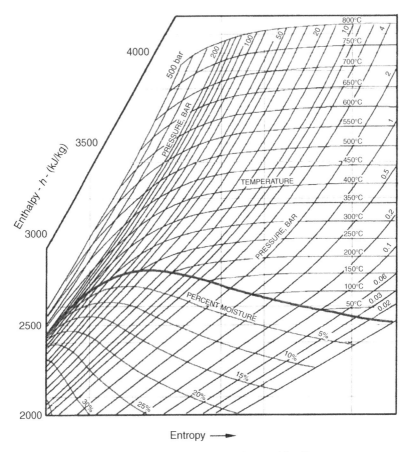

**Figure 14-3** *Steam Mollier diagram, SI units.*

## ENERGY WASTE IN CONDENSING STEAM TURBINES

The fluid cracking unit had a 5000-hp wet gas compressor driven by a steam turbine. I wanted to calculate the percent of the driver steam wasted due to the nonoptimum operation of turbine. The steam pressure available was 585 psig (600 psia) at 600°F. The cooling-water temperature available was 86°F.

I placed a pressure gauge on the steam chest. As the governor speed control valve was only open by 68%, the pressure in the steam chest was 135 psig (150 psia). Also, the two-stage jet system attached to the surface condenser was not in service. Only the single-stage "hogger jet," intended for startup, was in service.

With 86°F cooling water, the two-stage jet system should have produced a vacuum in the surface condenser of 51 mmHg (2 in. Hg positive pressure). But with only the hogging jet in use, the actual vacuum in the surface condenser was 150 mmHg (6 in. Hg or 3 psia).

Referring to the Mollier diagram (Figure 14-2), let's now consider two cases:

- *Ideal operation.* Enthalpy of 600 psia and 600°F steam = 1300 Btu/lb. Dropping down the chart on the constant-entropy line to 2 in. Hg yields an enthalpy of the exhaust steam of about 840 Btu/lb (best to use your own Mollier diagram, as I have dropped a little below my chart). Or the amount of work (expressed in thermal energy units) that may be extracted from each pound of steam is 1300 − 840 = 460 Btu/lb.
- *Actual operation.* Enthalpy of the motive steam is still 1300 Btu/lb. However, steam expansion starts at 150 psia and 1300 Btu/lb on the Mollier diagram. Dropping down to 6 in. Hg on the chart (3 psia) on the constant-entropy line yields an enthalpy for the exhaust steam of 1000 Btu/lb. Or, the amount of work that may be extracted from each pound of steam is 1300 − 1000 = 300 Btu/lb.

That is, the turbine should require only 300 ÷ 460, or 65%, of its current steam demand. The problem could, at least in part, be resolved by closing an additional hand (i.e., port or horsepower) valve on the steam chest. At the time the observations above were made, one of three of the hand valves was already closed, to keep the governor speed control valve in an operable position.

If you do not understand the above, I recommend the chapter pertaining to steam turbine operations in my book *Troubleshooting Process Operations* [3].

## COGENERATION PLANTS

I assumed several things in this preceding calculation:

- The steam turbine governor valve is wide open and has not created any $\Delta P$. This assumption can be made largely correct by manipulating the "port valves" on the steam chest. I have explained about the hand or port valves in my book, *Troubleshooting Process Plant Control* [2].
- I assumed 100% efficiency in converting the kinetic energy of the expanding steam to work. In reality, I should derate my calculated benefits by 10% for turbine and coupling inefficiencies.
- The steam bypass valve shown in Figure 14-1 is assumed to be closed. I can make this assumption pretty much true if I size my steam turbine port valves, when fully in service, to supply 100% of the required steam flow to the reboiler tubes. Again, this is best explained in reference [2].

I ask that you make a donation of 5% of the electric energy saved using my design to my Process Engineering Temple of New Orleans. Such contributions are tax deductible.

Unfortunately, I made the mistake of allowing my friend Mark to review this chapter. Mark observed, "Lieberman, I'm going to use your idea of replacing the parasitic control valves on the steam supply to my reboilers with topping turbines to drive pumps and simultaneously generate electric power. It's a good idea, but I'm not going to make any donations to your temple."

"But, Mark," I observed. "We are a new religious order."

"Anyway, Lieberman, this is the same idea you filed a patent for in 1977. Remember what the patent examiner told you then? Remember, he wrote you a report saying all this was 'open art' and that anyone skilled in such an art would already be aware of all the principles you claimed were so innovative in your patent disclosure."

"Gee, Mark, we've been friends for a long time. Over three decades."

"We are not friends," Mark fumed. "Anyway, what's the difference between your idea and the cogeneration plant you and I helped design in 1994 in Eagle Point, New Jersey?"

"Oh! That's a much bigger facility, but process-wise, it's kind of similar," I conceded.

"Not similar, Lieberman, it's the same concept. And how about the expander turbines that extract work from the catalytic cracker regenerator flue gas [1]? The turbines are used to drive the regenerator air blower, with the excess work generating electricity from the air blower's startup three-phase ac motor."

"Also rather similar to my innovative concept," I responded weakly.

"Yeah, Lieberman! And the roto-flow turbo expander in Kingsville, Texas. And the hydrocracker power recovery turbines at the Chalmette refinery and H-Oil Unit in Ft. McMurray, Alberta."

"Well, Mark. In the spirit of friendship, a small contribution to my Temple fund would still seem appropriate."

"Okay, Norm," Mark said more calmly. "It's a good idea to save electric power. Instead of contributions, how about a coffee and beignet in the French Quarter?"

"That sounds fair," I agreed. "Let's go."

## EXTRACTING WORK FROM REBOILER STEAM USING EXISTING EQUIPMENT

So Mark and I sat in the Café du Monde, across from Jackson Square, and watched the tourists clatter past as the horse-drawn carriages rattled across the cobblestones.

"Norm," Mark said, "doesn't your idea of the new turbine–motor–pump rotating assembly (Figure 14-1) violate one of the fundamental principles of your phony religion? You know, the commandment that says, 'Thou shall not purchase new process equipment.'"

"True. But I have a way around that problem," I explained. "Look, Mark, many of our existing pump services have both a motor and a steam turbine drive: a motor drive on one pump, a steam turbine on the other pump. We could use the existing turbine on any pump in the unit to extract free work from the reboiler steam supply, without buying any new turbines, pumps, motors, or even any new control valves.

"Sketch up your idea," Mark said, and handed me his coffee-stained napkin (see Figure 14-4). "I'm interested."

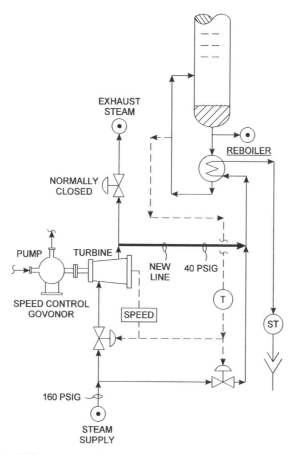

***Figure 14-4*** *Split-range control used to recover free work from steam supply to boiler.*

My idea was to use an existing pump and steam turbine to extract potential work from the steam going to the reboiler. Only a new line connecting the turbine exhaust to the reboiler steam supply line is required. The novel feature of Figure 14-4 is the control scheme. Split-range temperature control is required to control the reboiler process outlet temperature. Override speed control is required to control the speed of the turbine. It works like this:

1. The reboiler temperature controller (TRC) calls for more steam.
2. The governor steam speed control valve opens first, as far as it can, but limited by the steam required by the turbine to satisfy its speed set point.
3. The existing steam temperature control valve then opens, as required, to supply the rest of the steam needed for the reboiler heat balance.

For simplicity, I have assumed that the required reboiler steam flow is in excess of the turbine steam flow required. Should this not be the case, the split-range control would have to include the exhaust steam valve shown in Figure 14-4. Control of the existing TRC valve and the existing governor speed control valve, using the new split-range control, is all quite conventional (see my book *Troubleshooting Process Plant Control* [2]).

"That's a pretty neat idea, Norm," Mark said, as he carefully folded the napkin and put it in his shirt pocket. "Just a piece of 6-in. carbon steel pipe and reconfiguring some control loops on the panel. Of course, it's all subject to a "management of change" process review. It's a smart way to run a bunch of our pumps for free on our alky unit."

"Excuse me, Mark, may I have my sketch back?"

"No, Norm. I believe this is my napkin."

"Yeah! But it's my sketch. It's my idea. Could you at least make a token contribution to my Temple?" I protested.

"I just did," Mark answered. "You ate it."

So Mark and I walked up the levee and watched the turgid Mississippi River glide by. The tankers pushed upriver against the mighty current. The grain-loaded barges, guided by powerful tugs, slid downriver. The sunlight sparkled on the brown waters, as the runoff from a third of North America rushed past our feet.

## THERMODYNAMICS IN ACTION

"Norm," Mark asked, "I have a question for you. You, who claim to know all about the properties of flowing steam and thermodynamics."

"Yes, Mark. I was divinely inspired in 1980 in Texas City as to the science of heat (thermo) in motion (dynamics). What's the question?"

"I have a topping steam turbine. The motive steam is 200 psig, superheated to 600°F. The turbine is exhausting to a 60-psig steam header. If the turbine exhaust steam is getting hotter, is this a sign that the turbine efficiency is going up or down? Norm, by efficiency, I mean the amount of usable power I'm extracting from each pound of motive steam."

"Why do you ask, Mark? Is this a trick test question you missed in your thermo class at Texas A&M?"

"No. I just thought this could be a neat way to monitor changes in my steam turbine efficiency. Then I would know when to fix the governor or the jet nozzles or to clean the turbine blades," Mark answered. "But if you don't know, forget it."

"Look, Mark. Ever notice that when saturated, dry 100-psig steam at 330°F blows out of a steam leak, it's not all that hot?"

"Sure, Norm! But why is that?"

"Cause the steam blows out with lots of velocity. It's like an 'isoentropic expansion.' That is the heat in the steam, as well as the pressure of the steam, is converted to kinetic energy. The sensible heat in the steam is the temperature of the steam. Yes, the steam cools off. Yes, the heat content of the steam is reduced. But the energy content of the steam is preserved."

"Yeah, Norm. That's what the word *Thermodynamics* must mean, converting the thermal energy of the steam to the dynamic energy (velocity) of the steam. Heat is converted to kinetic energy."

"Right. And Mark, the more thermal energy that is converted to velocity, the more work that can then be extracted from each pound of steam in the turbine case," I explained.

"How does that work? I'm confused."

"Look, Mark. You've seen a windmill. The thing that makes the windmill turn is the wind: or the velocity of the air striking the sails of the windmill. A steam turbine is just the same. It's the velocity of the steam striking the turbine blades, that cause the turbine to spin. The faster the. . . ."

"Okay, I got it," Mark interrupted, "The faster the steam strikes the blades or buckets on the turbine wheel, the more work that can be extracted from each pound of steam. And as more thermal energy in the form of steam temperature is converted to more dynamic energy, in the form of steam velocity, then more work can be extracted from each pound of steam."

"Correct, Mark. And it follows then that a hotter turbine steam exhaust temperature indicates. . . ."

"Indicates, Mr. Lieberman, that the turbine efficiency is getting worse. I could even calculate my loss in turbine efficiency from your Mollier diagram (Figure 14-2)."

Mark stared across the river to Algier's Point. He seemed lost in thought. The brown wake from a passing barge crashed against the levee wall. "So, Norm. That's what the isoentropic expansion stuff is all about. It's expanding the steam so as to maximize the velocity of the steam as its pressure is reduced. Then the high-velocity steam, with lots of momentum, can be used to spin a wheel. Kind of like a little windmill."

"Mark, you're getting smarter by the minute. You may actually become a real engineer yet."

"You know something, Norm, I never really understood thermodynamics until just now. How come they didn't explain it to us at school like you explained it just now?"

"Maybe, Mark, because your professor at school didn't understand it either."

## STEAM TURBINE CHECKLIST

The blades on a steam turbine need to be cleaned periodically to remove the entrained hardness scale deposits. Degradation of the turbine efficiency may be monitored by tracking the turbine exhaust steam temperature as described above. Other than that, the following procedures should be followed to minimize the waste of the motive steam:

- Check that when the governor is in the wide-open position, the $\Delta P$ across the governor is not more than 10 psi or 5% of the supply steam pressure (whichever is smaller). Have the unit machinist reset the governor linkage to achieve this target.

- Close the process-side discharge-to-suction spill-back valve.
- With the spill-back valve shut, reduce the governor set speed until the down-stream process valve is in a mostly open but still controllable position.
- Close the hand valves (also called port or horsepower valves) connected to the steam chest until the governor speed control valve is in a mostly open but still controllable position. Note that for large steam turbines the Steam rack, will perform this same function automatically.
- If the turbine is exhausting to a surface condenser, minimize the condenser pressure. In New Orleans in the summer a vacuum of 26 in. Hg is good. A vacuum of 20 in. Hg is poor. Vacuum depends on elevation and the cooling-water supply temperature.
- When exhausting the steam to a steam header, keep the turbine exhaust pressure as close as possible to the steam header pressure.

Note that the actual supply steam pressure to the turbine is really the steam chest pressure. The actual exhaust steam pressure is the turbine case pressure.

Don't think I do this stuff everyday. I only went through this complete exercise once on all the turbines at the Texas City refinery during the 1980 strike at the amine and sulfur plant complex. But since I did it myself, with my own hands, I actually know what I'm writing about. To quote one of my more famous colleagues, Confucius says:

- "That when I listen, I forget.
- That when I see, I remember.
- That when I do, I understand."

So, my dear friend Mark, I believe that's why you and I did not understand thermodynamics at school. We were taught by instructors who didn't understand the subject themselves because they never practiced our trade, process engineering, with their own hands. Doing is the key to understanding.

## REFERENCES

1. National Snow and Ice Data Center Report, 2008.
2. Lieberman, N. P. *Troubleshooting Process Plant Control*, Wiley, Hoboken, NJ, 2008.
3. Lieberman, N. P. *Troubleshooting Process Operations*. 4th ed., PennWell Publications, Tulsa, OK, 2009.

# Chapter 15

# Expanding Compressor Capacity and Efficiency

My wife, Liz, asserts that my claims of being very smart are just a cover-up for my laziness. This is not true, as I am very smart. As an example, when a client calls with a complaint about a missing invoice number, I respond, "Sorry. But our computers are down." I can then spend the remainder of my day in peace and tranquility. It is not my fault that I don't have a computer. But if I had one, I'm sure it would be down.

Another one of my areas of intelligence is expanding compression capacity without purchasing any new compressors.

## RECIPROCATING COMPRESSORS

When I worked in Chicago, we assumed that reciprocating compressors (recips) had an efficiency of 90 to 95%, and that centrifugal compressors had an efficiency of 65% for small, slow-speed machines and 85% for more modern, very large, higher-speed machines.

In practice, vendor estimates for centrifugal compressor efficiency are correct if the wheels are clean. Vendor estimates of 95% efficiency for reciprocating compressors are pure fantasy. There are a number of problems that cause loss of recip efficiency:

1. Pulsation dampener plates
2. Valve velocity losses

*Process Engineering for a Small Planet: How to Reuse, Re-Purpose, and Retrofit Existing Process Equipment*, By Norman P. Lieberman
Copyright © 2010 John Wiley & Sons, Inc.

3. Broken valve plates
4. Broken valve springs
5. Leaking piston rings
6. Stuck valve unloaders
7. Pulsation losses

Fixing these problems is not just a matter of capacity, but also of energy efficiency. Very few of my clients run "indicator cards" or "*P–V*" diagrams on their reciprocating equipment. The indicator card will tell the maintenance folks which of the items 2 through 7 above are contributing to reduced compressor capacity and efficiency. The card is a plot of cylinder internal pressure vs. piston position. It is done electronically by an outside service company. All reciprocating compressors can be diagnosed using this technique as described in my book, *Troubleshooting Natural Gas Processing* [1]. Bloch, in *A Practical Guide to Compressor Technology* [2], refers to the indicator card as a *PV diagram*.

Have your maintenance department use the indicator card to determine if a valve has to be fixed. Sometimes, valves malfunction because the valve unloaders get stuck. If opening and closing the unloader valve does not affect compressor performance, that's a sign that the unloader is jammed. That is, the valve plate is jammed by the unloader fingers in an open position.

Pulsation losses can be reduced by reducing the valve spring tension. But then the valve will leak more. The indicator card will guide you as to optimum spring tension.

Incidentally, molecular weight does not particularly affect selection of the reciprocating compressor valves; the effect is only on valve velocity losses, which increase with gas density. But such losses account for only a small amount of reciprocating compressor inefficiency.

If a valve cap feels hot, it does not necessarily mean that particular valve is bad. One inlet or one outlet valve on that particular cylinder end is probably defective and all the valve caps are hotter because gas is not moving through the cylinder.

## PULSATION DAMPENER PLATES

My main point about recips relates to dampening vibrations due to pulsation. The maintenance staff in many plants use restriction orifice plates, called *pulsation dampeners*. The frequency of pulsation in recips is a function of the acoustical design of the piping and the speed of the machine. That cannot be altered by the process engineer or the maintenance department. However, the amplitude of the pulsations can be suppressed with these restriction orifice plates. This reduces damaging vibrations and hence maintenance costs. But the cost in energy and capacity paid for these plates may be far larger than the maintenance costs. To estimate the energy consumed for a motor-driven recip by the pulsation damping plates, use the formula

$$\Delta \text{ amps} \propto \left(\frac{P_2}{P_1}\right)^A - 1 \qquad (15\text{-}1)$$

$$A = \frac{K - 1}{K} \qquad (15\text{-}2)$$

where $K$ is the ratio of specific heats. Values for $K$ are:

- $H_2 = 1.41$
- $N_2 = 1.40$
- Air $= 1.40$
- Methane $C_1 = 1.31$
- Ethane $C_2 = 1.22$
- Propane $C_3 = 1.17$

where

- $P_2 = $ discharge pressure, psia
- $P_1 = $ suction pressure, psia

Next, proceed as follows:

*Step 1.* Measure the pressures upstream and downstream of the orifice plates at the suction and discharge of the orifice plates. You will now have four pressure readings.

*Step 2.* Using equations (15-1) (15-2), calculate $\Delta$ amps for the two pressure readings upstream and downstream of the plates.

*Step 3.* Repeat for the two readings inside the plates.

*Step 4.* Divide $\Delta$ amps in step 3 by $\Delta$ amps in step 2.

*Step 5.* Multiply the result from step 4 by the current amperage load on the motor. The difference between this calculated value and the current amperage load on the motor is the energy wasted by the pulsation dampener orifice plates.

To properly estimate the capacity reduction due to the orifice plates, you will need the vendor curves showing the variation in the gas capacities at different suction and discharge pressures. However, very roughly, the loss in capacity will be proportional to the reduction in the recip's absolute suction pressure.

In the same sense, try to minimize upstream and downstream pressure restrictions due to piping and valves. Reduction in gas temperature has a minor effect, in proportion to the percent decrease in degrees Rankine. For instance, reducing the gas from 100°F to 80°F will increase capacity by only 4%.

## CALCULATING RESTRICTION ORIFICE PLATE $\Delta P$

To calculate the pressure drop through your revised pulsation dampeners, in psi:

$$\Delta P = (3 \times 10^{-4})D_V V^2$$

where
   $D_V =$ vapor density, lb/ft$^3$
   $V =$ velocity, ft/sec

One should select a $\Delta P$ value that is a compromise between wasting electrical energy and maintenance costs. For example, you could pay the salary of a full-time maintenance person by saving 10% of the energy needed to drive a 1200-brake horsepower pipeline booster natural gas reciprocating compressor. Historically, management would never support such economics. But in our new environmental crisis, where we must factor in the $CO_2$ accumulation in the atmosphere, these relatively small economies in energy input must assume a new importance in our decision-making process.

## ADJUSTABLE HEAD-END UNLOADERS

One of the most environmentally friendly features of a reciprocating compressor, is the adjustable head-end unloading pocket. This is a large difficult-to-turn wheel at the end of the cylinder. Turning the wheel counterclockwise pulls back a plug inside the cylinder, away from the valve ports. This creates a bigger space inside the cylinder between the valve ports and the cylinder head. As a result, when the piston completes the end of its travel toward the cylinder head (this position being called *top dead center*), more gas is trapped between the piston head and the end of the cylinder. The trapped volume of gas expands as the piston moves back and away from the cylinder head. As this trapped gas expands, its pressure is reduced. The pressure of this old, trapped gas must fall to less than the inlet gas pressure before new gas will flow into the cylinder. Thus, turning the adjustable unloading pocket counterclockwise will reduce the amount of gas compressed per stroke. This reduction in capacity does not change the compressor's energy efficiency.

The technically correct way to express this concept is to say that we are reducing the volumetric efficiency of the compressor, but maintaining the adiabatic compression efficiency, by increasing the starting volumetric clearance.

If the reciprocating compressor is a variable-speed gas engine or steam turbine drive, I would just slow it down to reduce the volume of gas compressed. If the recip is a constant-speed compressor, there are two commonly used methods to reduce the volume of gas compressed:

- *The EVIL method*. Open the discharge-to-suction spill-back line. This is bad because the gas that is recirculated has to be recompressed. Also, the recirculated gas is hot and increases the temperature of the feed gas to the compressor. Hotter gas consumes more energy to compress. The increase in energy is proportional to the increase in the absolute temperature at the compressor suction.

• *The GOOD method.* Use the adjustable head-end unloader. Typically, the maximum amount of capacity reduction will be 25%. However, this does not waste any energy. Not all reciprocating compressors are equipped with such head-end unloading wheels. But almost all can be retrofitted with them. It's simple. It only requires unbolting the existing cylinder head and bolting on a new head-end unloader (which are, incidentally, quite expensive).

I was instructing a process seminar in Durban, South Africa not too long ago. An older and very smart operator, Chris, told the class he never used the head-end unloader on his reciprocating compressor. Rather, he and his co-workers just used the spill-back to reduce the reciprocating compressor's capacity.

"Mr. Lieberman, it's easier for the operator to open the 3-in. spill-back valve a bit than to turn that big head-end unloader valve."

"But, Chris," I objected, "That's wasting a lot of electric power on the compressor motor driver and promoting $CO_2$ emissions to make the wasted electricity. You're doing the devil's work."

"And just how, Saint Lieberman, was I supposed to know that? An action is evil only if one knows that it's wrong. None of our brilliant unit engineers ever explained the function of the adjustable head-end unloader to us humble operators," replied Chris.

There are many hundreds of simple ways to save fuel, steam, boiler feed water, and electric power on process units. But if the plant engineers do not explain these options to the hourly operators, it's we engineers who are acting against the future of our little planet.

## GAS-FIRED ENGINES

Most of the reciprocating compressors I've worked with in the Texas City refinery and in the gas fields in Webb County, Texas, were driven by engines fired with natural gas, which is largely methane. Methane is, on average, 23 times worse than $CO_2$ per mole as a greenhouse gas. If one cylinder of a 12-cylinder engine is not firing because the spark plug wire is broken or loose, 8% of the engine's fuel is vented un-burned to the atmosphere with the engine's exhaust. But which is the defective cylinder?

Here's an old trick from the gas fields in South Texas:

*Step 1.* Put the fuel gas supply valve in a fixed position, so as not to control the recip's speed automatically.

*Step 2.* Disconnect each spark plug wire, one at a time.

*Step 3.* Disconnecting the spark plug that does not reduce the compressor's speed is a sure indication of a defective firing cylinder.

I have written an entire book, *Troubleshooting Natural Gas Processing*, about such old tricks. The methods described were already archaic when I wrote that text in 1985. However, gas-fired engines and reciprocating gas compressors are even older.

## CENTRIFUGAL AND AXIAL COMPRESSORS

I was delivering a lecture to a group of young engineers about fractionation efficiency at the Spartan refinery in Louisiana when a crash of thunder stopped my presentation. The rain beat loudly against the corrugated metal roof. An hour after the storm had abated, Dick Lejuene, the FCU (fluid catalytic cracker unit) supervisor rushed into the classroom.

"Mr. Lieberman, can you come with me?" Mr. Lejuene asked.

"Okay, Mr. Lejuene. These guys aren't listening to me anyway. Is there a problem?"

"There sure is. I just lost 20% of the airflow to my catalyst regenerator. It's the storm. That rain blew into the blower's inlet filter. The filter's got soaked. I must have a big $\Delta P$ through the filter. Probably this is restricting the airflow. The motor amps are down, too. I reduced the cat feed rate by 20%. Mr. Babin wants you to check the filter pressure drop. We can change out the filter elements (see Figure 15-1). Mr. Babin says you'll decide."

I filled a plastic bottle with colored water. Using a section of plastic tubing, I checked the pressure drop across the filter (see Figure 15-2). The $\Delta P$ was 2 in. $H_2O$, compared to the 4 in. $H_2O$ $\Delta P$ that it had been last month. The rain had washed the dirt off the filter elements, thus reducing $\Delta P$. But where had the dirt gone to? Apparently, the dirt was now on the blades of the axial air compressor rotor (Figure 15-3).

"Dick, the rain has washed the dirt off the filter and into the air compressor. The rotor and stator blades are fouled," I explained.

"Mr. Lieberman, I don't agree. If the rotor was dirty, the compressor efficiency would be down and we would have pulled more amps, not less. Don't you know that it's harder to spin a dirty machine than a clean machine?"

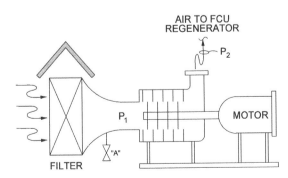

**Figure 15-1** *Dirty axial air compressor has a reduced amp load.*

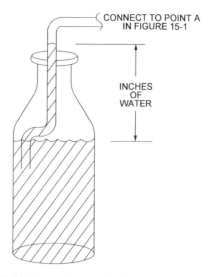

CONNECT TO POINT A
IN FIGURE 15-1

INCHES
OF
WATER

**Figure 15-2**   *Measuring inches of water pressure drop across a filter.*

"Sorry, Dick, but you are quite mistaken. Both for axial compressors, as well as for centrifugal compressor rotors (see Figure 15-4), the dirtier the rotor, the lower the amperage load on the fixed-speed motor driver. If we had a variable-speed gas turbine or steam turbine running at a constant shaft horsepower, the compressor would run faster as it fouled."

"You all ain't from around here, Mr. Lieberman. I guess you're from Chicago or New York or some such place. Around these parts we keep rotoring equipment clean, to make it run better," argued Dick.

"Okay, Dick. Let me apologize."

**Figure 15-3**   *Axial air compressor.*

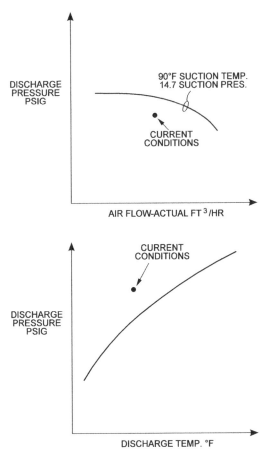

**Figure 15-4**  Simplified operating curves for centrifugal compressor.

"Oh, it's okay. Everybody makes mistakes. Even Mr. Babin. He hired you, didn't he?" Dick laughed.

"No, Dick. Let me apologize for not giving you a complete explanation."

"Go ahead, Lieberman. Mr. Babin—he says we all got to work with you."

"Look," I explained, "The air compressor suction pressure hasn't changed at $P_1$ and the discharge pressure is the same at $P_2$. So the compression ratio is constant" (see Figure 15-1).

"Okay. I got it. Go ahead."

"That means if that the efficiency of the air compressor was constant, we would need the same work or amps (amps being an expression of electrical work) for each million scf/hr of air moved."

"But, Lieberman, the airflows dropped off by 20%," said Lejeune.

"But the amps are only down 10%," I added. "This means that we're consuming 10% more work for each volume of air compressed.

"Yeah! Cause the rotor's dirty. But where's the extra amps gone to?"

"To heat! Look, Lejuene, before the rain, the compressor discharge temperature was 220°F. Now it's 235°F, even though it's been about 90°F ambient the entire day. The air's temperature rise is up by over 10%."

"Yeah! I guess the dirty blades on the rotor don't bite as hard into the air at the machine's suction. That does make sense."

"Well yes, you're right. Or we could say that the dirt on the rotor and stator blades causes the axial air compressor to run on an inferior capacity and efficiency curves. Or, it's like a shovel with mud stuck to it. You're not going to move as much soil from the ditch, but you'll also work harder, even though you're working slower and the ditch. . . ."

"Say, I ain't no ditch digger." Lejuene objected.

"It's just an analogy," I said.

"An ant-logy? Look, I guess you're just saying we got to wash the rotor, not change the filter elements," Dick Lejuene concluded.

## NEXT DAY

"Good morning, Dick, how's the air compressor? Cleaned up any?"

"We water-washed it. Everything's back to normal. Mr. Babin, he said to say thanks. So, thanks."

"You know, this would be a good opportunity to check the air compressor's performance. Right after you cleaned it," I suggested.

"Look, Mr. Lieberman, I'm real busy. I just said it's back to normal. But go do your New York engineering stuff if that's what you want. Mr. Babin, he said we got to work with you."

So I then plotted on the compressor's performance curves (see Figure 15-4) the current discharge pressure, flow, and discharge temperature. Since the molecular weight of air is constant and I assumed that the suction pressure of the compressor is constant (a good assumption along the Gulf coast unless there's a hurricane) and the suction temperature is constant (not always a good assumption), I can call the vertical axis the discharge pressure. Really, it should be "polytropic feet of head."

Regardless, Dick Lejeune's current operation, plotted on the curves shown in Figure 15-4, indicated that his "normal" operation was almost midway between the design curves and yesterday's fouled rotor operation which I observed right after the rain storm. I reviewed my calculations with Craig Babin, the technical services manager. I explained that they were still losing about 15% of the air capacity and wasting about 8% of the motor amps or electric energy. Most likely, the compressor blades were still partially fouled.

Months later my theory was proven correct. The axial air compressor was shut down and cleaned according to GE's (i.e., the manufacturer's) instructions. Upon startup, the air compressor ran properly both as to head vs. capacity and head vs. power curves.

**Figure 15-5** *Centrifugal wet gas compressor.*

## CENTRIFUGAL COMPRESSOR

Figure 15-5 shows the four wheels of a centrifugal compressor which looks similar to the three natural gas compressors that I had at the Laredo, Texas, dehydration station. These rotors would foul every six months. As the rotors fouled, the gas flow would be reduced. However, the turbine was a natural gas–fired driver, running at its maximum driver horsepower. Thus, as the rotor fouled with salt and a thick greasy substance, the compressor spun faster. This wasted some of the energy used to drive the turbines. Also, so much capacity was lost that I had to run all three compressors to move the required flow of natural gas.

But not all four wheels were equally dirty. The first wheel was clean. The second wheel was slightly fouled. The third wheel was terribly clogged with black greasy gook and dirty white salt. The fourth wheel was dirty, too, but not as bad as the third wheel.

Why was the first wheel so clean? I recalled my instructions from Mr. John Housman, my old mentor at Amoco Oil: "Norman, keep the wheels wet. Spray heavy gasoline into the compressor suction to keep the wet gas from drying out. The heat of compression causes the gas to get hotter and dry out on the wheels. Spray in enough gasoline so you push the calculated dry-out point out past the last wheel. Based on the:

- Temperature
- Pressure
- Gas composition
- Gasoline composition used as a spray
- Compression ratio
- $K$ (i.e., $c_p/c_v$) of the wet gas

calculate how much of the gasoline spray you need so as not to reach dew point conditions inside the compressor. Then all the wheels will stay wet. Just remember to use a misting type of spray nozzle at the inlet to the compressor," concluded Mr. Housman.

So that's what I did in Laredo. The fouling rate dropped. I moved more natural gas with less turbine fuel. But best of all, I could move the required amount of natural gas with just two rather than three compressors. This reduced the overall $CO_2$ emissions from the Laredo station by about 30%.

## CLEANING THE CENTRIFUGAL COMPRESSOR ROTOR

The story about the axial air compressor at the Spartan refinery did not have a happy ending. Dick Lejeuene phoned me one day at my home in New Orleans.

"Well, we cleaned the air compressor. It works a lot better now, Mr. Lieberman."

"Thanks for calling, Dick. Glad I could help out. But I bet that's not why you called," I said.

"Didn't you tell us to also clean the centrifugal wet gas compressor?"

"No, I did not," I answered.

"Well, we did. Mr. Babin said that's what you told him. You gonna call Mr. Babin a liar?"

"Okay, Dick. Just tell me the problem."

"We cleaned the wet gas compressor rotor. We lifted the top of the compressor case and pulled out the rotor. Just like you predicted, the rotor was all gunked-up. We cleaned the stationary elements, too. When we put it back together, the compressor ran real good. Right on the manufacturer's head vs. flow curve. Right on the head versus. . . ."

I interrupted, "Right on the manufacturer's head vs. power curve. And you didn't need all that. . . ."

Lejeune now interrupted, "Yeah, Lieberman. You know that my wet gas compressor is a constant-speed motor-driven machine. After we cleaned it, like you told us to do, it developed lots more pressure and lots more flow. Which meant lots more work. The work came from the motor's amps. I didn't need all that extra flow and I didn't want any extra differential pressure. So I throttled back on the control valve (see Figure 15-6) at the compressor suction. But then the compressor started to surge. So, to stop the surge I opened the discharge to suction spill-back. Then the high-motor-amp load alarm lit up on the compressor. So I had to close back on the spill-back valve. Then the compressor started to surge. So I opened the suction throttle valve a little and the high-amp light came back on the motor, so I. . . ."

"Okay, Dick. I got it."

"Yeah, Lieberman. Now thanks to you, I'm running with my wet gas compressor close to its surge point, and close to the motor tripping-off on high amperage. Before you had us clean the rotor, the spill-back valve (Figure 15-6) was shut and the suction throttle valve was open. Now both valves are half-open and the motor's electric use has jumped by 15%. I didn't need any more $\Delta P$ or flow. Why did you have us clean that [expletive deleted by publisher] rotor?"

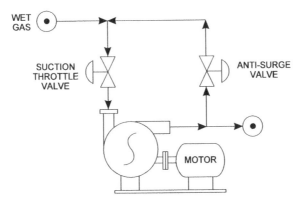

**Figure 15-6** *Suction throttle valve is closed to reduce power consumed on the motor.*

"Listen," I shouted into the phone, "I didn't tell you people to clean the wet gas compressor. Obviously, if you clean the wheels and the stator, you're going to move more gas, which requires more power. And that's not going to be completely offset by the improved adiabatic compression efficiency of. . . ."

"Adi Batic who?" asked Lejuene.

"Look, Dick. It's like cleaning the mud off your shovel when you're digging a ditch. The cleaner shovel will move more soil but. . . ."

"Lieberman, I ain't no ditch digger," Dick Lejuene shouted and hung up.

The moral of this story is that the compressor was designed for a lower molecular weight than current operating conditions. Thus, when clean, it produced too much head and flow with the higher-molecular-weight gas. One solution to this problem is not to keep the rotor dirty and working at a low efficiency, but to remove one of the four wheels shown in Figure 15-5. Mechanically, this is a complex and expensive modification, but a modification that will save electric power on the fixed-speed motor driver.

This story also illustrates a major benefit for having variable-speed drivers on centrifugal wet gas compressors rather than fixed-speed motor drives, especially when dealing with a variable-molecular-weight wet gas. Perhaps the existing motor could have been modified to a variable-speed driver by use of frequency control. Then the speed could have been reduced as the molecular weight increased.

The reason that throttling on the suction of a centrifugal compressor reduces the amp load on the motor driver is a result of the shape of the compressor performance curve. Referring to Figure 15-4, note that as we move toward the left, flow is dropping faster than the discharge pressure is rising. Thus, when we suction throttle, a relatively small increase in the compression ratio results in a relatively big reduction in the flow. It is this disproportionate reduction in flow that mainly results in the reduced motor amps.

From what I have just said, you may think that suction throttling may actually increase the amp load on the motor driver if we move very far to the right on

Figure 15-4. Note that as I slide down this curve to the right, it is getting progressively steeper. At some point the increased $\Delta P$ required due to suction throttling will cause only a small reduction in flow. If I kept sliding down this curve, by raising the compressor suction pressure or lowering the compressor discharge pressure, the flow could not increase further. This is called stonewalling or, more correctly, choke flow.

One characteristic of operating near choke flow is that suction throttling does not noticeably reduce the amperage load of the motor driver, because the ACFM stays constant regardless of the head changes.

This was all explained to me by Heinz Block [2].

## REFERENCES

1. Lieberman, N. P. *Troubleshooting Natural Gas Processing: Wellhead to Transmission,* PannWell Publications, Tulsa, OK, 1987. Reprinted by Lieberman Books, 2008.
2. Bloch, H. P. *A Practical Guide to Compressor Technology,* 2nd ed., Wiley, Hoboken, NJ, 2006.

# Chapter 16

# Vapor–Liquid Separator Entrainment Problems

I was driving west on I-10. I had a project at the Jupiter refinery in Baytown and was running late. Just past the Lake Charles exit, I noticed an old Ford pickup with a flat tire. A young-looking unshaven guy was sitting in the grass near the truck. I pulled over.

"What's up?" I asked, as he stood up and walked over to my car.

"Mister. I got me a flat and no spare." He was dressed in old, worn-out coveralls and dirty sneakers, and had unkempt bushy hair.

"Do you want me to call a tow truck? You belong to Triple A?" I asked.

"Naw. I got me no money. Anyway, I is just bout outta gas. This ole truck's finished. Alternator's stopped working, too. But thanks for stopping, Mister."

"So what are you going to do?"

"Don't rightly know. Nothin, I guess," he answered, as he sat back down in the weedy grass.

"Look. Get in and I'll take you to the next gas station. Let's get some breakfast. Okay?"

"Thanks," he answered, as he sat next to me.

"I guess you're not working."

"No, sir. I done lost me my job. Was workin' at that there refinery in Norco. But they done gone and shut down and I got my ass laied off. Was workin' as an assistant operator on that there gas oil hydrotreater. I had me a nice little house down by the river, too."

*Process Engineering for a Small Planet: How to Reuse, Re-Purpose, and Retrofit Existing Process Equipment.* By Norman P. Lieberman
Copyright © 2010 John Wiley & Sons, Inc.

"Would you believe it," I said in surprise," I'm on my way to a job to troubleshoot a problem on a gas oil hydrotreater at the Jupiter refinery."

"What's their problem, mister? By the way, my name's Leroy. What's yours?"

"Norm. The problem Leroy is that the recycle gas, reactor effluent vertical separator vessel is carrying over. Leroy, you know what entrainment means?"

"Say Mister! You ain't that there Norm Lieberman guy, is you? That dude who writes bout troubleshootin and stuff?"

"That's me," I said, both amazed and flattered.

"Sure's I knows what that there entrainment is. That's when that foamy liquid pukes over, even when the level's okay. Those Jupiter boys! They should have had them a horizontal, not a vertical separator. That'll slow down them vapors. It's them vapors moving too fast that entrains droplets. Guess them Jupiter cats ain't none too smart."

"Look, I got an idea. Come with me to the Jupiter plant now. I need someone to help me field check pressures and temperatures. You know how to connect up pressure gauges and change out dial thermometers? I'll pay you $60 for today."

"Cash money?" Leroy asked. "Man. I is ready."

## LEVEL PROBLEMS IN V-603

Leroy stood next to the separator in my spare boots and extra Nomex and studied the liquid level in the gauge glass in V-603 (Figure 16-1).

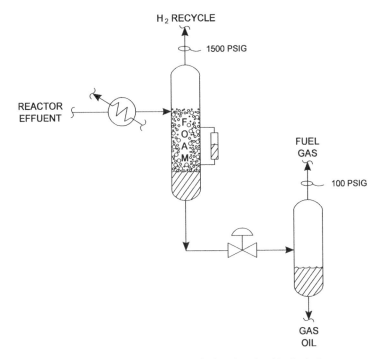

*Figure 16-1* *Foam level is higher than level indicated.*

"You all know somethin, Mr. Norm, that there level is too high."

I had made a mistake bringing Leroy into the Jupiter plant. Mr. Anderson, the operations manager, did not approve of Leroy.

"Lieberman," he said, "Your company has been retained to provide a detailed vessel sketch for a new and larger replacement vessel for V-603, the reactor effluent high-pressure vapor–liquid separator. We have budgeted $800,000 for the new 1800-psig-rated vessel. We do not appreciate your associate, Mr. Leroy Johnson, lecturing us that we should have used a horizontal separation drum, with a neutron backscatter, level detection system to deal with foam."

"Speaking about Leroy Johnson, I bill his time out at $1000 a day. Is that alright?" I asked.

"Fine! Just see that he fills out his contractor time sheet correctly at the end of the week. But kindly advise him to stop telling our hourly operators that our engineers don't understand about level control. The level in the separator is not the problem. The vessel is too small and needs replacement. I have budgeted $3,200,000 for the design, purchase, and installation of this new vessel. Our engineering headquarters division in Virginia has already approved of the replacement."

"Okay, I'll tell Leroy. Sorry about it."

"Of course, we at Jupiter are an equal opportunity employer and are very proactive in promoting minority employment," Mr. Anderson added.

"So are we," I agreed. "That's why I actively recruited Leroy myself for our staff this year."

## FOAM-INDUCED CARRYOVER

"You see, Mr. Norm, that there level you all see in the gauge glass is not the real level in V-603. The real level's higher. Probably up almost to the feed nozzle," said Leroy. "Look closer, Mr. Norm, at the level glass. See that liquid kind of dribbling back through the top tap and running down the glass. That's foam. The foam drains through the top tap. Foam breaks down when it touches lots of surface area, like the gauge glass. That's the liquid part of the foam you see draining through the glass. The top tap of the glass is only 1 foot below the feed nozzle. If the foam gets itself above the feed nozzle, the vapor in the feed blows that foam into the hydrogen recycle gas. I done seen it all before in Norco. Man, I sure do miss my little house on River Road down by the levee."

"Why all the foam?" I asked. There was something odd about Leroy. Something that I hadn't noticed before.

"Well, Mr. Norm. At high pressures and low temperatures, hydrogen and heavier liquid hydrocarbons like gas oil create a very stable foam. Wouldn't happen with jet fuel or diesel. Wouldn't happen at lower hydrogen pressures. Wouldn't happen at high temperatures. Has something to do with surface tensions."

"What else, Leroy?" I asked with growing suspicion.

"Mr. Norm, the other problem is particulates. Like iron sulfides. You all see that them Jupiter cats are bypassing the hydrogen feed caustic scrubber. That leaves them chlorides in the hydrogen, which makes hydrochloric acid in the reactor. That causes corrosion, which makes iron sulfide particulates. Particulates cause foam. Like dirty water foams-up worse than clean water. You should tell them Jupiter boys to put that there caustic scrubber back online."

"But listen, Leroy. How can you be so sure the problem is a high foam level? You don't have x-ray vision, do you?"

"I'll prove it," Leroy answered. He called over to the outside operator, "Kevin, tell Petey to cut that level in V-603 by 10% on the panel."

I watched the level in the glass drop by 8 in. But more to the point, the liquid flow dribbling through the top tap stopped completely.

"You understand now, Mr. Norm?"

"No, Leroy, what's happening here?"

"I'll explain," Leroy said. "When the foam in the vessel rises above the top tap of the gauge glass, the level in the glass does not represent the level in the vessel. The level in the glass represents the density of the foam in the vessel, in proportion to the density of the liquid in the glass. As foam density is just a small fraction of liquid density, the foam level in the vessel must always be higher than the external liquid level indicated in either the gauge glass or the level-trol."

"So the level set point is too high," I concluded.

"Correct, Mr. Norm," said Leroy.

"And if we get the level too low?" I asked. "What's going to happen then?"

"Then the flow of fuel gas (Figure 16-1) will jump up as the foam layer slips out through the bottoms LRC. Mr. Norman Lieberman, perhaps I should summarize:

*Step 1.* Set the LRC to hold a level below that which causes liquid to dribble through the top of the glass, but above the level that causes an increase in fuel gas flow.

*Step 2.* Put the NaOH scrubber back into service to keep the HCl out of the hydrogen feed gas to the reactor.

*Step 3.* Inject a silicon defoamer into the separator feed (to the first vessel in Figure 16-1) at a concentration of 5 ppm to suppress carryover, but only on an emergency, backup basis.

*Step 4.* Install a Nuetron backscatter level radiation detector [sold by K-Ray] to monitor the true foam density and foam levels.

[I noticed a growing radiance behind Mr. Leroy Johnson. His bushy hair glowed softly in the golden light. A smile spread across his youthful brown face.]

*Step 5.* Cancel the V-603 replacement vessel project. Mr. Norm. It's just a waste."

And then a gust of wind blew through the Jupiter refinery. A cloud of dust rose and obscured V-603. And when the air cleared, Leroy was gone. As I drove home the next day, I looked for his truck on the west-bound side of I-10. The truck, too, was gone. Most unusual for the Parish Highway Department to be so prompt in removing an abandoned vehicle [1].

In conclusion, horizontal vapor–liquid separators, not vertical separators, should be used in this service. The horizontal vessel will have a larger liquid surface area, for the same liquid residence time, than that of the vertical separator. The rate of foam dissipation is proportional to the vapor–liquid interface surface area in a vessel.

## ENHANCING DEENTRAINMENT RATES

Entrainment velocity, $V_a$, in ft/sec, is defined as

$$V_a = K \left( \frac{D_L}{D_V} \right)^{1/2} \tag{16-1}$$

where
  $D_L$ = liquid density (note: units cancel out)
  $D_V$ = vapor density (note: units cancel out)
  $V_a$ = vertical component of velocity, ft/sec

$K$ is selected as follows:

- Low entrainment relying only on gravity settling: $K = 0.10$
- Moderate entrainment, relying only on gravity settling: $K = 0.2$
- High entrainment, relying only on gravity settling: $K = 0.3$
- Excessive entrainment, relying only on gravity settling: $K = 0.4$
- Moderate entrainment using a demister pad: $K = 0.25$ to $0.30$
- Moderate entrainment using a demister pad and a vapor distribution device (such as a vapor horn): $K = 0.30$ to $0.35$

A *demister* is a woven metal mesh screen, typically 4 to 8 in. thick. Smaller droplets of liquid impinge on the metal mesh and coalesce into larger droplets, which drip off the pad. If the demister is used in a fouling service, or if the demister corrodes, the pad will partially plug. This will lead to higher localized velocities and thus the demister pad will make entrainment worse, not better. Thus, in many refinery applications, demister pads should not be used. If you do use a demister pad, select a metallurgy for the mesh that is not subject to corrosion. Monel, Hasteloy, titanium, and 317 stainless steel are some possible choices (see Chapter 13 for details).

For a demister to work, the vertical velocity using equation (16-1) must be in the range calculated using a $K$ factor between 0.20 and 0.3. Demisters are ineffective at much lower velocities.

## VAPOR DISTRIBUTION AS AN AID TO DEENTRAINMENT

The purpose of this chapter is to provide you with design advice so as to avoid the purchase of new vessels used as vapor–liquid separators. One such method is to dissipate the momentum at the inlet nozzle of the feed by converting a single radiant entry to a dual tangential entry, using the vertical baffle configuration described in Chapter 8. Check the section of Chapter 8 on retrofitting of the Texaco Marine vacuum tower for a sketch of this baffle. If the inlet nozzle velocities are less than 70 ft/sec. I would not be concerned with any inlet vapor distribution. If inlet nozzle velocities are greater than 150 ft/sec. I would be quite concerned with the even distribution of the vapor feed.

An additional method to improve the vapor distribution is to place a chimney tray distributor above the inlet nozzle, as detailed in Figure 16-2. The chimneys may be either round or rectangular. The area of the chimneys should be calculated using a

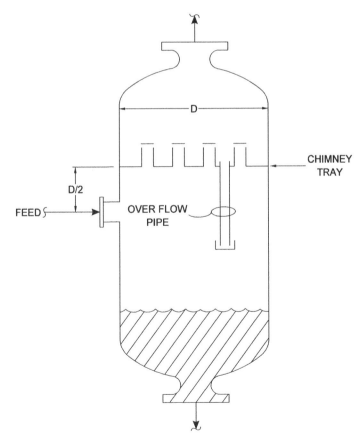

***Figure 16-2*** *Chimney tray used as a vapor distributor to promote deentrainment.*

velocity so as to produce a pressure drop of a few inches of liquid:

$$\Delta P = (0.8)\frac{D_V}{D_L}V^2 \qquad (16\text{-}2)$$

where

$D_V, D_L$ = as defined in equation (16-1)

$V$ = vertical velocity of vapor flowing through chimneys, ft/sec

$\Delta P$ = pressure drop of the vapor, inches of liquid at $D_L$

Don't forget the overflow pipe. Otherwise, the deentrained liquid will overflow the chimneys and be reentrained, thus defeating the purpose of the chimney tray distributor. The overflow pipe should have a separate seal pot, as shown in Figure 16-2; or, extend the overflow pipe down to the liquid at the bottom of the vessel. If the overflow pipe is not sealed in some way, vapor will blow up the pipe. Then liquid will be unable to drain down the overflow pipe, countercurrently against the rising vapor flow. This would then defeat the whole purpose of the overflow pipe, and flooding could result.

## REFERENCE

1. Adapted from Leo Tolstoy's short story, " What Men Live By."

# Chapter 17

# Retrofitting Shell-and-Tube Heat Exchangers for Greater Efficiency

Lazarus, resurrected from the grave by Jesus, had lain in his shroud for three days. Cold and dead, beneath the rocky soil of Israel. Afterward, he saw things differently than the rest of humankind.

He looked at a young tree flowering in the spring and saw it fallen and decayed after centuries of struggle against the storms of Judea.

He looked at the young bride, and saw her old and bent, a hag broken by the years of childbirth and hardship.

He looked at the land, flowing with milk and honey, and the newly turned moist earth and saw the hillsides eroded and desiccated, every tree cut by human hands. That which had been a garden turned into a desert.

Why Lazarus was granted this foresight is not known to us.

In 1974 I produced 1,200,000 gallons per day of high-octane gasoline at my alkylation unit in Texas City. I looked on this as a service to society. Under my supervision, 0.7% of all the gasoline consumed in the United States was refined from lighter hydrocarbon components. I was fulfilling a socially valuable task. Now I have been granted a different insight into the production of jet fuel and automotive gasoline. Is this really a socially valuable task or an ultimately destructive activity? And why have I been granted such insight?

Money makes the world go round. Financial liquidity is the lifeblood of our business economy. Refining of crude serves the same function in the equally important energy economy. We reach out to the far ends of the Earth, to the remote corners

*Process Engineering for a Small Planet: How to Reuse, Re-Purpose, and Retrofit Existing Process Equipment,* By Norman P. Lieberman
Copyright © 2010 John Wiley & Sons, Inc.

of our planet, and draw nature's bounty into Baytown and Beaumont and Baton Rouge, to distribute to the citizens of our land gasoline and jet fuel. But for what purpose? Why does each family in our nation need 4 gallons of these products every day? Where must this all end? Like Lazarus, I can see the end: New Orleans submerged; the desiccated farmlands of the Midwest; the offshore rigs abandoned, having exhausted their reservoirs. Why has this foresight been granted to me by a power beyond our comprehension? I guess to share knowledge as to how to avoid such an end. And the bit of knowledge that I would now like to relate pertains to improving the performance of shell-and-tube heat exchangers, which is a strategy to reduce fossil fuel consumption in refining and petrochemical process plants.

## NEW HEAT EXCHANGER DEVELOPMENTS

In the past few decades a number of novel designs have been promoted to enhance exchanger performance that reuse the exchanger shell and process piping:

- Helical baffles
- Twisted tube bundles
- Tube inserts in the form of springs
- Various turbulators inside the tubes
- Sintered metal tube coating
- Low fin tubes
- Spiral heat exchangers

Rather like Lazarus, I look at these developments and see discarded tube bundles that testify to ideas that had little practical value. My vision is not based on insight, but on experience and discussions with my clients who have tried these new features and have, in general, been disappointed. The main problems encountered are:

- Springs break apart
- Cannot be cleaned
- Alignment of components not designed correctly
- Not mechanically robust
- Tube inserts get stuck
- Worked well when new, but did not stand up in service
- Promotes fouling on tube exterior
- Promotes erosion at points of contact
- Promotes tube sheet roll leaks

My wife, Liz, has written a complete description about these new developments in another text[1]. But if such retrofits are doomed to failure, what methods are available to improve heat exchanger performance? Let me use as an example the service that most of us are familiar with: preheating cold crude oil with clean hot products or pump-arounds. We will assume that crude is on the tube side and that we are upstream of the desalter and hence dealing with dirty, salty crude oil. The heating fluid is thus assumed to be on the exchanger shell side. The first suggestion I have is the best documented of all, and it pertains to processing recovered slop oil.

## RERUNNING SLOP OIL

Recovered refinery slops and waste oil should never be rerun on the crude unit. These slops are will be contaminated with:

- *Cracked materials*. Especially from a delayed coker on visbreaker, and to a lesser extent from the FCU (fluid catalytic cracking unit). These cracked hydrocarbons contain diolefins.
- *Oxygen*. Especially oil recovered from sewer skimming operations, from coker blowdown systems, and from cone roof storage tanks.

When diolefins are heated to 300° to 350°F, or come into contact with hot tubes above this temperature in the presence of air, free radicals are formed. The free radicals polymerize to form gums. These gums will glue particulate matter to the surface of the tubes. Rapid loss of the heat transfer coefficient results. At a refinery in New Orleans, the heat transfer coefficients on crude preheat exchangers would drop by an order of magnitude, to less than 15 Btu/(hr-ft²-°F) in less than one year. I have attributed this decline to their practice of running off slops along with virgin crude oil.

At a refinery in Texas City, slops were runoff by charging them to the overhead vapor line quench of the delayed coker. Or, one could use the slop as quench oil to the effluent from your visbreaker soaker drum. Or, inject the slops into the riser on the FCU. All these methods will destroy the diolefins before they can polymerize and gum-up the inside of a heat exchanger.

- *Floating suction in crude tank*. For many years, a major refiner's R&D personnel tried to develop a chemical additive to inhibit exchanger fouling on the crude preheat exchanger train. All for naught. Eventually, some unsung hero from the R&D division discovered a truly effective antifoulant program: a crude tank floating suction, as shown in Figure 17-1. Potential fouling deposits settled out in the crude charge tank. To take advantage of this concept, we found that it was necessary not to run the tank mixers shown in Figure 17-1. The rate of sludge accumulation in the crude storage tanks increased and sludge removal and disposal were also increasing problems. Further, intermediate pipeline crude storage tanks that supplied crude charge to the refinery would routinely turn on their tank mixers to flush their sludge into our pipeline crude supply. Sudden losses in our heat transfer coefficients were directly correlated with these upstream activities, which were then curtailed.

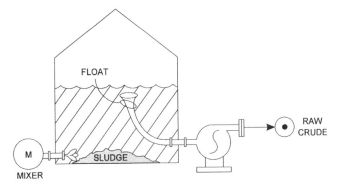

*Figure 17-1*   *Crude tank floating suction used to reduce heat exchange fouling.*

- *Exchanger online spalling.* Jerry Edwards, my goat-herding colleague, and I discovered this effective technique quite by accident. One day we inadvertently opened an exchanger's crude side bypass valve and closed the crude inlet valve (Figure 17-2). The hot shell-side pump-around was left flowing. The tubes then heated to the pump-around temperature of 300°F. Within 15 minutes we had restored normal operations. But in this 15-minute interval, two favorable results had accidentally been achieved:

  1. The tube-side $\Delta P$ value had gone down by approximately one-half.

  2. The exchanger heat transfer coefficient had magically increased from about 20 Btu/(hr-ft$^2$-°F) to 45 Btu/(hr-ft$^2$-°F)!

Thereafter, Jerry and I tried this procedure on all our exchangers upstream of the desalter, with comparably favorable results. I imagine that the sudden heating of the tubes melted off the deposits. Or, more likely, the thermal expansion and contraction of the tubes loosened the fouling deposits from the inner surface of the tubes. When

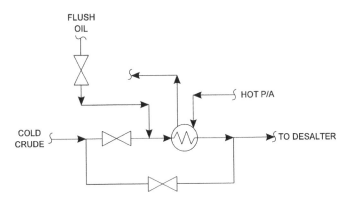

*Figure 17-2*   *Online heat exchanger cleaning by heat spalling.*

we restored the crude flow, we flushed these deposits out of the exchangers and into the downstream desalters.

Years later I noted that one of my clients, operating a visbreaker, was using the same procedure. However, they also used a small flush oil flow (as shown in Figure 17-2) to dissolve out heavier organic fouling deposits. The flush oil consisted of heavy aromatics from their reformate product naphtha splitter bottoms.

> • *Feed interruptions.* Many of my clients have observed that erratic operations of their crude units (i.e., frequent short shutdowns and restarts) have resulted in a temporary improvement in crude preheat. My exchanger online spalling technique, described above, is just a variant of this rather common experience. Caution! For equipment downstream of the desalter, thermal cycling may cause hazardous flange leaks to occur on the crude side of the heat exchangers. The crude may be above its autoignition temperature.

I ask the reader to contrast my techniques with the recommended approaches by various vendors to combat tube-side fouling:

- Tube inserts to promote turbulence
- Spinning springs to erode off deposits
- Chemical fouling dispersants
- Antioxidants to retard free-radical polymerization

If you try these vendor-promoted applications on your unit, after enough time and experience have been gained, you may develop the same insight and negative expectations that I have toward these new and expensive proprietary techniques, intended to reduce the fouling rate in crude preheat exchanger service.

## TUBE-SIDE VELOCITY AND METALLURGY

Other techniques to reduce exchanger fouling are more costly. One method is to increase the number of the tube-side passes. Going from two to four tube-side passes will double the tube-side velocity. For many types of crude, a reasonable minimum tube-side velocity is 5 to 6 ft/sec. Note that doubling the tube passes increases $\Delta P$ by a factor of 8, due to doubling the velocity and doubling the flow path length.

Another option is to retube the carbon steel bundle with nine chrome tubes. The objective is to reduce tube surface pitting and roughness due to corrosion. The smoother the heat transfer surface remains, the fewer the fouling deposits that will adhere to the surface. Nine percent chrome tubes can be used in direct physical contact with the existing carbon steel tube sheets and carbon steel tube support baffles, without concern regarding corrosion due to galvanic action. Neither 304 or 316 stainless steel can be used in crude preheat service, due to the potential for chloride stress corrosion cracking.

Note that retubing a carbon steel bundle with 316 stainless steel will promote galvanic corrosion to both the carbon steel tube support baffles and the carbon steel tubesheet. Use of a sacrificial anode will retard such corrosion but is not good design practice, as often no one remembers to renew the anode. If you note that corrosion to the carbon steel component is continuing and the anode's size is not diminished, the anode is not wired up correctly. I just learned this yesterday from a Pakistani engineer in one of my seminars.

## SHELL-SIDE SEAL STRIPS

Another very simple, essentially no cost, environmentally friendly way to increase the heat transfer coefficients is by the use of seal strips. These are shown in Figure 17-3. They work in pairs to reduce shell-side bypassing of the tube bundle. You will find that for exchangers with liquid on the shell side, seal strips are already present on most tube bundles. The problem is that most of my clients are not aware of the importance of these simple components and thus do not maintain them. Especially if the shell-side fluid has an appreciable viscosity (i.e., above 10 cS), the seal strips will significantly improve the overall heat transfer coefficient.

Sometimes, seal strips are also used along the sides of the bundle if the exchangers are equipped with horizontal cut baffles. Dummy tubes are also an effective way to block off potential shell-side bypass areas. Placement of such dummy tubes costs very little. They are not attached to either tube sheet. Best to inspect your tube bundle after it has been pulled and visually locate where such dummy tubes can be placed, so as to minimize the potential for flow shell-side bypassing [2].

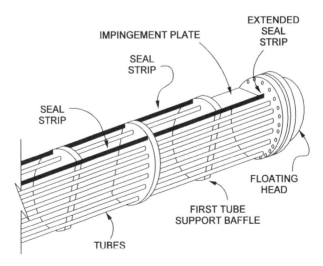

**Figure 17-3**   *Seal strip tube bundle.*

## HIGH-VISCOSITY FLUIDS

I was the tech service manager at the Ever Hopeful refinery in Norco, Louisiana in 1982, when the evil Abram Engineering Company installed eight large banks of shell-and-tube heat exchangers to cool the plant's large industrial fuel oil product. Industrial fuel oil has a very high viscosity at ambient temperatures. When products are cooled so that their viscosity is over 10 or 20 cS (50 to 100 SSU), cooling becomes difficult due to laminar flow. If higher-viscosity material is flowing through the tubes, the laminar flow creates a high-heat-transfer film resistance.

The Abrams Engineering Company had placed the industrial fuel oil inside the tubes for one reason: It is easier to clean the tube side of an exchanger than the shell side. They then placed the cooling water on the shell side. Unfortunately, the heavy fuel oil at the laminar boundary layer inside the tubes developed a very high viscosity (probably over 100 cP), and thus severely retarded cooling. The result of this fiasco was that:

- The heat transfer coefficient observed was 8 Btu/(hr-ft$^2$-°F), compared to the design of 40!
- The fuel oil run-down temperature was 310°F instead of the 190°F designed.
- The captain of the fuel oil barge refused to load our fuel oil on his vessel, as it was 100°F too hot.

Mr. Stanley, the general manager of the Ever Hopeful refinery, directed me to invite the barge captain out for a few drinks and then introduce him to some ladies of the evening down in the French Quarter in New Orleans. Meanwhile, the hot fuel oil would be loaded on his barge. But I had a better idea. It was to switch the shell-and-tube side flows.

You see, there is no such thing as laminar flow on the shell side. Although the Reynolds number may be low (i.e., less than 2100), due to high viscosity, laminar flow cannot develop on the shell side, due to vortex shedding.

As liquids flow perpendicularly across a tube, swirls are created. You may see such swirls as a river flows across an old tree stump that is sticking up in the current. These swirls create turbulence in the form of vortex shedding, which prevents a laminar boundary layer from retarding heat transfer.

"But, Norm," objected Mr. Stanley, "how can we clean the shell side of such an exchanger if the high-viscosity fouling material is on the shell side?"

I explained that this particular tube bundle could be cleaned readily on the shell side because of the geometry of the tube bundle:

- The tube pitch arrangement was a rotated square rather than a triangular pitch. Triangular pitch bundles do not have cleaning channels between the tubes.
- The space between the tube walls was $\frac{1}{2}$ in. rather than the more standard space of $\frac{1}{4}$ in.

- The tubes themselves would be more rigid, as they were 1-in.-O.D. tubes rather than the more standard $\frac{3}{4}$-in.-O.D. tubes.

So I retrofitted the eight banks of shell-and-tube exchangers by placing the cooling water on the tube side and the industrial fuel oil on the shell side. The heat transfer coefficient observed rose to 40 Btu/(hr-ft²-°F), and the heavy industrial fuel oil rundown temperature dropped to 200°F. Again, the improvement was due to promoting shell-side vortex shedding. Incidentally, my experience with placing a high-viscosity fluid on the shell side to promote vortex shedding is based on the shell-side cross-flow velocity, exceeding 3 ft/sec, for vortex shedding to develop. By *cross-flow velocity*, I mean the flow between the tubes and the tube support baffles, as calculated at the edge of the tube support baffle.

All this took a few weeks to implement. In the meantime, we ran down much of the hot fuel oil product to fixed roof tanks. One of the tanks had a heel of water. The water eventually flashed to steam from the 300°F fuel oil. The roof then sailed off across Norco, Louisiana, and landed in a neighbor's yard. But King Zeus was watching out over the Ever Hopeful refinery, and no one was killed.

The efficiency of many heat exchangers that cool viscous products can be improved by placing the viscous product on the shell side. Unfortunately, the tube support baffle spacing may have to be adjusted to keep the cross-flow velocity above 3 ft/sec. This is mechanically difficult and expensive to do.

## WATER COOLER FOULING

One easy way to get more cooling capacity out of an existing shell-and-tube cooling-water exchanger is to reduce the number of tube passes. For example, one should change from four-pass to two-pass, presuming the cooling water flow is on the tube side. Stop! Do not email me to say that I've contradicted what I wrote in the section "Tube-Side Velocity and Metallurgy." Let me explain.

- The tube-side pressure drop for the circulating cooler water is fixed. It's the difference between the cooling-water supply pressure and the header pipe return pressure.
- When I convert a tube bundle from four-pass to two-pass, the tube-side $\Delta P$ must remain constant.
- However, now the water has to flow only half as far with the two-pass tube bundle configuration as it did in the four-pass configuration.
- Thus, the water will flow faster, and we will also have more water flow, as the number of passes is reduced.
- Faster water flow keeps the muck and the biological deposits from sticking to the inside walls of the tubes. More water flow reduces the cooling-water outlet temperature and thus reduces the rate of carbonate fouling deposits inside the tubes.

Note that changing from four-pass to two-pass is mechanically complex. A new, central channel head pass partition baffle is required, which must fit in a new groove in the channel head tubesheet.

## VAPOR EVOLUTION IN COOLING WATER

"Lord of the Universe, my dear friend Chago Leo, operations manager of the ESSO LARGO Aruba refinery, died of liver cancer in 1999. Chago once asked me why, when he wanted to maximize water flow to his old seawater-cooled condensers, he often had to throttle-back on the cooling-water outlet valve. Master of Creation, I pray that you pass on the following observations to Chago:

- If the sea-water rises to too high above grade, especially as it gets warmer, the air dissolved in seawater flashes out of solution. A single ounce of air will evolve into a cubic foot of vapor (i.e., air expands as it comes out of solution). The evolved air weighing a single ounce will occupy the same volume as will 64 lb of seawater. Holding a very slight backpressure at the cooling-water outlet isolation gate valve suppresses the evolution of this dissolved air and thus permits more seawater flow through the tubes.

- Of greater importance are tube leaks in light hydrocarbon cooling service. If a lighter hydrocarbon leaks into the cooling-water final tube pass, it can flash to a vapor and displace some of the cooling tower circulating water, in the same way that evolved air can displace seawater. Throttling the cooling-water outlet valve would then partially suppress the tube leak and result in more cooling-water flow to the condenser, and more efficient cooling. My point is that this is also an indication of light hydrocarbons (and in the Aruba refinery, lots of $H_2S$ also) leaking into the cooling tower return header. These hydrocarbons will then escape into the atmosphere. As you all in Heaven well know, of course, light hydrocarbons accumulating in the atmosphere are a major contributor to greenhouse gases. Methane, for instance, now accounts for about 10% of the current global warming trend on the surface of our small planet Earth. Praise the Lord! Amen."

By the way, did you know that in most process plants, the major cause of loss of cooling-water heat exchanger capacity and efficiency is tube-side fouling? And that the major cause of such fouling is biological in nature: that is, algae and bacteria? If you don't believe me, climb up and look at the cooling-water distribution decks on your cooling towers. They will probably be covered with a thick layer of algae. The food source for all of this organic growth are the hydrocarbon tube leaks inside your process heat exchangers. I guess this is nature's way of combating global warming.

In 1974–1977, I spent huge amounts of money with the chemical vendors, buying biocides and chlorine to combat organic fouling of my cooling-water system in Texas City, all of which ultimately contributed to environmental degradation. Dear reader, don't be like I was then. Find and fix the hydrocarbon leaks in your water coolers.

Once a week, crack open the high-point vents on your cooling-water lines. Next, using a portable gas test meter, check for hydrocarbons or $H_2S$. Don't follow my evil example of waiting until I had 4000 bsd of isobutane blowing out the top of my alkylation unit cooling tower in Texas City before I finally shut down the alky depropanizer to fix a 4-in. hole in the floating head in the depropanizer's overhead condenser tube bundle.

Actually, the only reason I shut down to fix that giant butane leak was because my boss, Frank Citek, noted that my unit's isobutane material balance was off. I couldn't have cared less about the environmental effects of the missing isobutane. I guess you can now see why Zeus and Hades have taken such a very special interest in my career as a chemical engineer.

## REFERENCES

1. Lieberman, E. T., and Lieberman, N. P. *A Working Guide to Process Equipment*, 3rd ed., McGraw-Hill, New York, 2008.
2. Gilmore, G. H. "No Fooling–No Fouling," *Chemical Engineering Progress,* Vol. 61, No. 7, July 1965, pp. 50–56.

# Chapter 18

# Reducing Sulfur and Hydrocarbon Emissions

The island of Carib is doing more than any place else in the world to combat global warming. I will explain. The reason we have not already been overwhelmed by global warming is due to two factors:

1. The ability of the ocean to absorb heat from the atmosphere.
2. The reflection of sunlight by sulfates in the atmosphere.

The first factor is temporary, and in the long run, self-defeating. The heat stored in the ocean, sooner rather than later, will heat up the atmosphere. The second factor, $SO_2$ emissions, causes acid rain, respiratory problems, and "global dimming." But how does this relate to Carib Island and their gigantic contribution in combating global warming by promoting global dimming?

## THE ERTC CONFERENCE

I had been invited to make a presentation at the European Refining Technology Conference in London on the subject of low-sulfur diesel and gasoline technology. Unfortunately, my presentation was not well received. To be honest, I was asked to leave the conference.

*Process Engineering for a Small Planet: How to Reuse, Re-Purpose, and Retrofit Existing Process Equipment,* By Norman P. Lieberman
Copyright © 2010 John Wiley & Sons, Inc.

"Ladies and gentlemen," I began, "this conference pertaining to reducing the sulfur content of diesel oil and automotive fuels in European refineries is inane and a waste of time. Don't you all realize that the island of Carib, with a population of 65,000 people, emits more sulfur dioxide and sulfur trioxide in one month than all of the Western European emissions from gasoline and diesel in an entire year? Don't you realize they are burning about 16,000 bsd of 4% sulfur fuel oil at the Pluto refinery and in the island's desalinization and power plants? Also, that neither of their sulfur plants is in operation as I speak. Also, that I. . . ."

"Mr. Lieberman," interrupted the conference chairman, "your remarks about gasoline desulfurization are quite out of order. Gasoline must be desulfurized to protect the catalytic converters used to suppress unburned hydrocarbon emissions from cars. The reduction in sulfur oxides is just a secondary benefit. We in Europe are quite concerned about automobile exhaust hydrocarbon emissions."

"Excuse me, Mr. Chairman," a young man with flaming red hair shouted from the back of the room. "It is you, sir, and the rest of this silly conference that are out of order. What you fail to understand is that most of our hydrocarbon emissions from internal combustion engines do not come from four-cycle modern automotive engines, but from two-cycle engines. Don't you realize that someone running a chain saw for a few hours emits more hydrocarbons than a new efficient compact car emits in a year? Our problem is not so much with cars, but with lawn mowers, boats, off-road motorcycles, and all the rest of our gadgets that use two-cycle engines. Unlike four-cycle engines, hydrocarbon vapors just blow right through such engines. If industry was really serious about curtailing smog and hydrocarbon emissions, they would stop selling hedge trimmers with two-cycle engines." [*Author's note:* An engine that requires you to mix oil with gasoline is a two-cycle engine.]

"Yes, Mr. Chairman, I have always trimmed my hedges in New Orleans with hand shears and my lawn mower is an electric. . . ." But at this point, before I could complete my presentation, I was ejected from the ERTC conference.

## SULFUR EMISSIONS FROM CARIB ISLAND

The gentle and steady trade winds always blow from Carib Island, across the blue Caribbean, toward Punto Frio in Venezuela. Good thing, too. If the wind ever reverses, it will choke the tourists sunning on the beaches with toxic $SO_2$ and $SO_3$ vapors. A dense, compact, and narrow white stream stretches from the stacks at the end of Carib Island to the shores of Venezuela. The airplane from Las Piedras, to the airport in Carib, flies right above this white path. As my flight descended below the long $SO_2$ cloud, shortly after the ERTC conference, I resolved to do something to reduce the sulfur emissions from the refinery. About half of the sulfur came from the combustion of heavy fuel oil on the island. That I could not influence. But the other half came from recovered hydrogen sulfide that was flared because of the lack of sulfur plant capacity. Or, to be more accurate, because neither one of the two sulfur plants was normally in operation.

The fundamental reason for the poor online factor for the two sulfur recovery plants were that no one particularly cared whether or not they ran. It is true that when neither plant ran, about 120 mt/day (metric tons per day) of hydrogen sulfide gas was flared. The flare burned with a giant beautiful blue flame at night—quite an attraction for the honeymoon couples on the other end of the island. The sales value of the lost sulfur production of about 110 mt/day was a pretty small incentive to operate and maintain these facilities. It was all an unfortunate carryover from the so-called "Jugo" days, meaning when Jupiter Oil owned and operated the refinery: in the evil days, before Jupiter Oil became an environmentally responsible corporate citizen.

I complained to my very good friend, the refinery manager, Mr. Haynes, about the sulfur plants. When I reminded him that I was representing Mother Earth and that it was her will that the $H_2S$ flare be extinguished, Mr. Haynes said, "Lieberman, don't come into my office, without an appointment, and start complaining about Ray Brinkly. If you want to speak to me, make an appointment with Linda. And I certainly don't want to hear your crazy 'Mother Earth' stuff again. If you're really an agent of Mother Earth, go ahead and figure out how to get those old sulfur recovery facilities up and running. Now get out of my office. I have my own problems."

With Mr. Haynes's kind encouragement, I went to talk to the desulfurization area outside operator. He explained that both sulfur plants were idled because of the same malfunction. The final condenser had a massive boiler feedwater (BFW) tube leak (see Figure 18-1). The 160-psig BFW flowed through the tube leak into the shell side, where it flashed to steam at 215°F and solidified the sulfur in the No. 3 condenser vapor outlet, which then plugged off.

The operator explained that two new condensers were on order and that their delivery was expected within the year, although when they would be installed was anybody's guess. Actually, he said, it was all a waste of time and money. The existing condensers had failed after a year's use, and the new ones would, he imagined, last no longer. Anyway, flaring the $H_2S$ didn't cost anything and had no adverse effects on the refinery's operations.

"And how about the environment?" I asked.

"Oh!" he answered. "The smoke all drifts away somewhere else. Over the ocean. It doesn't matter. We've been flaring for years. Even in the old Jugo days."

## SULFUR CONDENSER MODIFICATIONS

I had two objectives in modifying the No. 3 condenser:

1. I wanted to use facilities already in place.
2. I wanted to complete the job in a few days, not a few years.

The problem was that a relatively small hole in a single tube in the final No. 3 condenser would cause a huge amount of high-pressure water to flood into the low-pressure sulfur plant tail gas. But how much waste heat was being recovered in this

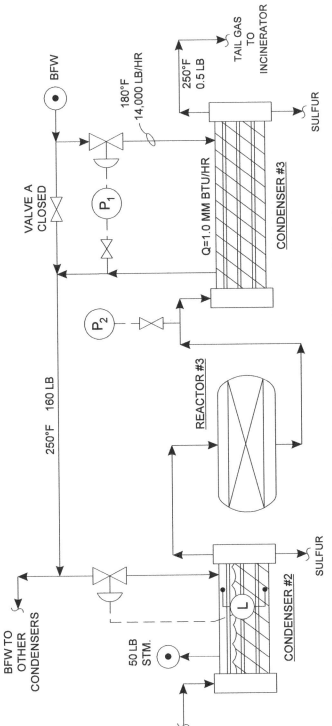

*Figure 18-1   Sulfur paint No. 3 condenser used to preheat boiler feedwater.*

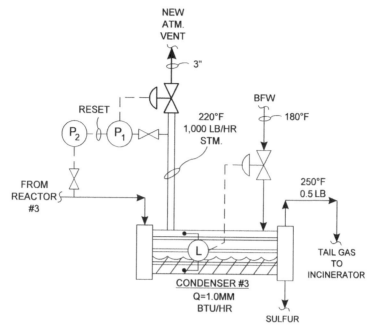

***Figure 18-2*** *Modifications to condenser.*

condenser anyway? It was not generating any steam. Just preheating the BFW from 180°F to 250°F. Was this important?

My idea was to bypass the high-pressure BFW around the No. 3 condenser entirely. I could vent the steam produced in the No. 3 condenser to the atmosphere. After all, the tail gas pressure on the tube side was essentially atmospheric pressure. Figure 18-2 represents my design submitted to Mr. Haynes, the plant manager. Valve A, shown in Figure 18-1, would be opened to supply cooler BFW to the other condensers. About 1000 lb/hr of steam would be lost, due to not preheating the BFW in the No. 3 condenser. The shell-and-tube-side pressures of the condenser ($P_1$ and $P_2$) would be kept about equal by resetting $P_1$ to make it equal $P_2$. The only new equipment needed would be the 3-in. atmospheric vent on top of the No. 3 condenser.

The water level in the condenser shell would be adjusted to keep the tail gas outlet temperature to the incinerator at about 250°F.

In reality, only the 3-in. atmospheric vent steam line was installed. The operators just manually dribbled in a few gpm of BFW and didn't pay any attention to the level. If the tail gas was getting too cold, they just shut off the BFW for a few hours. The new 3-in. line vented with no backpressure control to the atmosphere. If a little tail gas or water blew out with the steam, it didn't matter.

I told the operators that what they were doing was okay for now, until I got Mr. Haynes to authorize the instrumentation and controls shown in Figure 18-2. But Linda, Mr. Haynes's secretary, never gave me an appointment to see him again. And the No. 3 condenser puffed along happily for years, with its own little atmospheric vent but without my clever controls. Unfortunately, the blue romantic $H_2S$ flame dimmed over the island. These old sulfur plants are not 20 ft from the sea. In all the 80-odd trips I made to this "One Happy Island," I never missed a chance to look out over the blue Caribbean Sea with both the No. 3 condensers blowing whispers of steam behind me.

## FINDING HYDROCARBON LEAKS IN SEAWATER COOLING SYSTEMS

One day as I looked across the azure waters, I noticed that the sea seemed to be boiling in the area of the cooling-water outflow. Also, what looked like a heat haze was rising from the sea, but only in the same small localized area.

"Oh, Mr. Lieberman, that's just a little gas bubbling out of the cooling water return," explained an operator. "It happens all the time. The gas just drifts away somewhere else. Over the ocean. It doesn't matter. It happened in the old Jugo days also."

On Carib Island, and in many other seaside process plants, especially outside the United States and Europe, the ocean is used as a supply for cooling water. If a condenser tube starts leaking, light hydrocarbons such as propane and butane are lost into the sea breezes. How can an operator detect when such an exchanger is leaking? Not too easily!

Ordinarily, finding high-pressure light hydrocarbons leaking from the shell side into the cooling-water tube side is quite easy. A high-point vent is opened on the cooling-water return header. If gassy water squirts out, you have found a tube leak.

The problem with seawater exchangers is that the water outlet is under vacuum. The same problem also applies for water coolers at an elevation above the cooling tower distribution decks. If one opens valve B in Figure 18-3, air will be drawn into the valve but nothing will vent out, because the water at that elevation is under a vacuum. To test for a hydrocarbon leak into the seawater return header, the following procedure is necessary:

- As shown in Figure 18-3, a $\frac{1}{2}$-in. line or hose should be extended from the top of the exchanger to grade. Immerse the end of this line in a bucket of water.
- Partially close valve A until water overflows the bucket. Note that the evolution of vapor from the bucket does not necessarily indicate a leak, as air will also be evolved from the warm seawater.
- Check the vapor escaping from the bucket with a gas test meter [lower explosion limit (LEL)] for hydrocarbons and for hydrogen sulfide. (*Caution:* $H_2S$ levels can be dangerously high, above 1000 ppm.)

I never had the occasion to find hydrocarbon liquid leaks only in this manner, but I suppose it would work just as well. If you only blow gas but no water overflows out of your bucket, a major leak is indicated in the tube bundle.

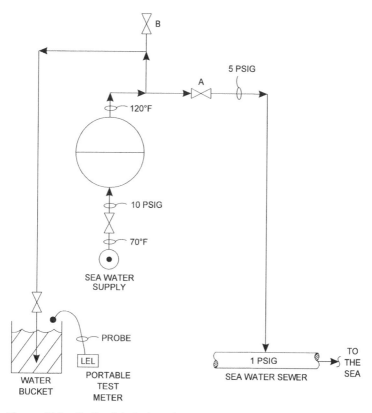

*Figure 18-3*  *Finding light hydrocarbon leaks in seawater cooling exchangers.*

## THE DRAGON

Ray Brinkly, the sulfur plant operating engineer, always warned me, "Norman, my friend, be careful of the Dragon."

The Dragon, Ray explained, was the sulfur plant tail gas incinerator. The purpose of the incinerator was to combust residual $H_2S$, $CS_2$, COS, and entrained liquid sulfur to $SO_2$. Also, to get the sulfur plant tail gas hot enough so that it would flow up the incinerator stack and drift away to Venezuela, as shown in Figure 18-4.

If the incinerator was not hot enough, the stack would not develop sufficient draft. The toxic $H_2S$-laden tail gas might then flow out of the air intake at ground level. In that case, the fuel gas flow was increased, to increase the stack temperature, so as to develop a larger draft. Also, the air intake damper would have to be opened at the side of the incinerator, to increase the combustion airflow. The incinerator was connected to the stack by a 100-ft-long, 6-ft-diameter duct. The original stack had fallen over during a fire many years ago. The incinerator had then been connected to an old, existing stack by the unusually long 100-ft duct.

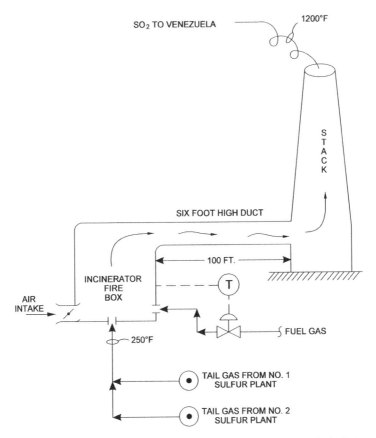

*Figure 18-4* Sulfur plant incinerator overpressured from air leaks in duct.

The problem was that the pressure drop through this duct was excessive. The duct was too long. Increasing the flue gas flow to the stack by 10% increased the duct's pressure drop by 20%. This could cause the pressure in the incinerator to increase above atmospheric pressure. Fire would then surge briefly out from the incinerator's air intake. Rather like your everyday fire-breathing dragon, I suppose.

When either No. 1 or No. 2 sulfur plants were in service, the $\Delta P$ in the incinerator line was not a problem and the Dragon slept. But placing both sulfur trains in service increased the duct's $\Delta P$ by a factor of 4. That is, pressure drop varies with line velocity squared. This was certain to awaken the Dragon's wrath. Even I, your fearless author, was frightened by the Dragon's breath. The sulfur plant operators, aware of their own mortality, were terrified. And my fixing the BFW leakage problems in the No. 3 condensers, which allowed both sulfur trains to operate all the time, greatly increased their fear of the Dragon, especially after one of the shift operators was badly burned.

Then, Mr. Haynes, the plant manager, shut down both sulfur trains and sent me the following email: "Lieberman, this is totally your fault. You were warned by

Mr. Brinkly about this potential safety hazard. Fix the problem at once or get off the Island. Also, you have not been invited to the annual refinery picnic on Sunday. Linda sent you the invitation in error."

## LOSS OF DRAFT DUE TO AIR LEAKS

So, while everyone else was at the silly company picnic, I crawled through the duct shown in Figure 18-4. The duct was made of carbon steel and had 3-in.-thick internal insulation. I was looking for internal obstructions inside the duct that might be contributing to the high $\Delta P$. But I could plainly see that there were no such obstructions. I didn't even have to use my flashlight. The bright Caribbean sunshine illuminated the interior of the duct. Especially where the short sections of duct were joined together or made turns, shafts of sunlight penetrated the gloom in dozens of spots.

When in service, the duct was under a vacuum of several inches of water (i.e., a positive draft). Any holes in the duct's skin would draw cold air into the duct. The cool air would increase the incinerator pressure in two ways:

1. Reduce the stack temperature and thus reduce the draft that was developed by the stack by about 8% for each loss of 100°F.
2. Increase the flow in the duct. The resulting higher velocities would increase the pressure drop in the duct.

After the picnic—which I didn't want to attend anyway—was over, the maintenance people reinsulated the defective portions of the duct. They also repaired the leaky carbon steel skin, especially the expansion joints. When the sulfur complex was restreamed, both sulfur trains could be operated without awakening the Dragon, which slumbers still.

And thus for many a month the beautiful blue light of the hydrogen sulfide flare was extinguished over the honeymoon paradise of the "One Happy Island" of Carib—until they again burned down the incinerator stack. But that's another story for another day.

## GLOBAL SULFUR EMISSIONS

One could ask: If sulfur emissions suppress global warming by increasing global dimming, are sulfur emissions good or bad? I guess sulfur emissions must on balance be very bad, because $SO_2$ and $SO_3$ result in acid rain. The acid rain then kills trees. Anyway, let's get back to Carib Island. I can't imagine that the current owners of the refinery are still flaring $H_2S$. [*Author's Note:* As of November 2009, the current owners of the refinery have shut down the plant more or less permanently.] But I would also guess that the Carib electric power and seawater desalinization plant is still consuming perhaps 8000 bsd of high-sulfur industrial fuel oil and thus emitting roughly 40 mt/day of sulfur in boiler plant flue gas.

It is all kind of irrational. For example, Hess Oil in New Jersey, at a great caustic consumption, is scrubbing refinery fuel gas to reduce its sulfur content below 0.01 mt/day. Yet the Mars refinery, just 20 miles from my home in Louisiana, is burning vacuum tower tail gas in their vacuum heater, with 2 mt/day of sulfur, because they are permitted to do so.

An ordinary three-stage sulfur recovery (SRU) plant will typically have a $97\frac{1}{2}\%$ recovery efficiency of sulfur. Emissions from the tail gas incinerator of a 140-mt/day SRU plant would then be $3\frac{1}{2}$ mt/day. This is bad, and environmental regulations would now dictate that a tail gas conversion unit (typically, a Shell-Scot type of unit) be installed to increase recovery to 99.9%. This will permit the sulfur in the incinerated tail gas to be reduced from $3\frac{1}{2}$ mt/day, down to 0.2 mt/day, which is good for the environment. However, these tail gas units require the reheating and then the cooling of the tail gas before incineration. This consumes lots of energy and generates more $CO_2$. But I suppose the reduction in acid rain is worth the extra energy consumption and the construction of new process facilities.

Yet at the same time, the same refinery that may have spent $80 million on a Shell-Scot tail gas unit for their sulfur recovery unit is producing marine diesel and bunker fuel for oceangoing ships with a 2% sulfur content. This fuel, when consumed, might emit 20 to 30 mt/day of sulfur. I mean, don't these ships sail around oceans on the same planet as the refinery?

Our problem is that we live on a really small planet and the sulfur and hydrocarbons emissions from Caribbean islands do not just drift off into outer space. When I worked in the 1960s in Whiting, Indiana, we used to say, "The solution for pollution is dilution." I now find that this is not quite true.

We, as process personnel, need to take a global view of our work. Burning 5% sulfur content visbreaker, vacuum tower bottoms residue, in the Carib refinery crude unit while people in Europe are hydro-desulfurizing FCU LCO (diesel) down to 100 ppm sulfur does not represent a rational global solution to environmental problems. If you doubt my qualification to comment on this critical subject of sulfur emissions, you are quite wrong. I am extremely qualified. When I was the technical manager of the Good Hope refinery in Norco, Louisiana, we were the largest producer of high-sulfur industrial fuel oil (i.e., bunker C) and marine diesel in the entire world. Maybe that's another reason that Mother Earth and Queen Hera have chosen me for this mission.

## EPILOGUE

My youngest daughter, Irene, has corrected this chapter for syntax and had a comment.

"Dad! I guess the parts about the boiling ocean and the Dragon and the leaking condensers are true. I know that your boring technical stories are based on real incidents. But the stuff about that ERTC London meeting? You must have made that up. Dad, you can't mean that engineers sit around and talk about our environmental problems without doing anything to correct them? Don't they realize that we live on a small planet with bad things floating around in the air? They should do something to stop these sulfur and carbon dioxide and hydrocarbon emissions. Not sit around in air-conditioned meeting rooms, watching PowerPoint presentations and eating fried shrimp. Dad, I think you made that part up. Like when you claim to be on a mission from Pallas Athena and the Creator."

No, Irene! I did not make up the story about the ERTC conference. It happened. Also, we are all agents of the Creator's will. I believe that the will of the Creator flows out of my hands and through my mind. While I'm dedicated to results, other engineers confuse talk for action.

# Chapter 19

# Hydrocarbon Leaks to the Environment

A typical refinery loses about 0.4 wt% of its crude supply to the flare, evaporation, theft, or fugitive emissions. The only two times that I investigated such losses myself, they were closer to 0.7 wt%. I recall that an Exxon refinery reported values in the range of 0.2 wt%, which represents an excellent operation. But these calculations are very complicated. The inputs to the refinery are not just crude oil. Natural gas, purchased slops, natural gasoline, additives, and such must also be included in the inputs.

The outputs are not just the products sold. They should also include:

- FCU catalytic coke
- Refinery fuel gas
- Sulfur sales
- Nitrogen converted to ammonia
- Red oils in spent sulfuric acid
- Transfers to integrated chemical plants
- Hydrogen converted to steam in the sulfur plants (i.e., moisture in the tail gas)

My point is that actual refinery losses are to some extent a function of the assumptions used in their calculation as well as actual losses. So even if your plant claims low losses of 0.2 to 0.3 wt%, there may still be lots of scope for improvement.

*Process Engineering for a Small Planet: How to Reuse, Re-Purpose, and Retrofit Existing Process Equipment,* By Norman P. Lieberman
Copyright © 2010 John Wiley & Sons, Inc.

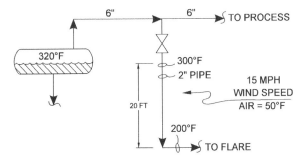

***Figure 19-1*** *Measuring losses to flare through a leaking valve.*

## MEASURING LEAKS THROUGH VALVES

I was reading the "Teachings of Buddha" last night. The Enlightened One's dying words were, "Rely Upon Yourself." These were his last words to his five disciples. We may apply Buddha's teachings to measuring leaks through valves. Let's say that we have a leaking valve to the flare, as shown in Figure 19-1. My objective is to estimate the flow rate through the leaking valve. I plan to use the following formula for this purpose:

$$Q = U A \, \Delta T \qquad\qquad (19\text{-}1)$$

where
$Q$ = heat loss in the 20-ft length of 2-in. pipe shown in Figure 19-1, Btu/hr
$U$ = heat transfer coefficient, Btu/hr-ft$^2$-°F
$A$ = exterior surface area of the section of 20-ft pipe, ft$^2$
$\Delta T$ = temperature difference between the average surface of the pipe and the ambient air, °F

$U$ for a bare pipe is a function of the pipe's surface temperature and the wind speed. For a cool pipe (100 to 200°F) and a low wind (0 to 10 mph), $U = 1$. For a very hot pipe (400 to 600°F) and a substantial wind speed (10 to 20 mph), $U = 4$. The $U$ value for this example is about 2.

$A$ for this example is about 13 ft$^2$. Don't forget to use the exterior area of the pipe, not the pipe I.D. (i.e., 2 in.).

$\Delta T$ is the temperature difference (average) between the pipe skin temperature (250°F) and the air temperature (50°F). For this example, $\Delta T$ is about 200°F. Our calculated $Q$ is then roughly

$$Q = (2)\,(13)\,(200) = 5200 \text{ Btu/hr}$$

The specific heat of most hydrocarbon vapors is about 0.6 Btu/(lb-°F). Therefore, the mass flow of vapor passing through the leaking valve shown in Figure 19-1 is

$$\text{leak} = \frac{5200}{(300°\text{F} - 200°\text{F})(0.6)} = 87 \text{ lb/hr}$$

The term $(300°\text{F} - 200°\text{F})(0.6)$ is the 60 Btu/lb of heat lost per pound of flow. This calculation ignores the conduction of heat along the pipe, pipe surface condition, condensation, and vaporization. Still, we can conclude that there is a hydrocarbon loss to the atmosphere of 60 to 100 lb/hr, or about 1 ton/day. But how to stop the leak?

## FIXING LEAKING VALVES

Overtightening a leaking gate valve will make it leak worse. Fluid will start to pass across the top of the gate. Leaks through most smaller valves can be stopped not by overtightening but, rather, by the following:

- Drill a hole ($\frac{1}{4}$ in.) through a packing gland just upstream of a plug, ball, or globe valve. For a gate valve, drill the $\frac{1}{4}$-in. hole under the seat of the gate.
- Inject a fibrous material such as wood chips (in the bad old days, I used asbestos fibers) or small plastic beads through the packing gland valve. A really nice material to use for many services is ordinary ion-exchange resin beads from water softeners.

Of course, the next time you use the valve, you will have to reseal the leak, so this is just a temporary fix. But, according to my Buddhist beliefs, all of life is transitory anyway, except that knowledge which we learn in any particular existence.

## DETECTING A LEAKING RELIEF VALVE

Pressure safety relief valves frequently leak in hydrocarbon service. If light liquid hydrocarbons are leaking past the relief valve, ice will form on the exterior of the pipe, downstream of the relief valve. Such relief valves will normally receive lots of attention. To identify leaking relief valves in vapor service, which are more common than those in liquid service, is a more difficult task. Most gases have a positive Joule–Thompson expansion coefficient, which means that they cool upon expansion. One example that is typical is water. A leaking steam relief valve on a 125-psig 350°F saturated steam header, leaking through the valve into a 5-psig header, would cool across the relief valve to approximately 320°F. Refinery fuel gas at 70 psig and 105°F leaking into the flare header would cool by about 10 to 15°F, to approximately 90°F. I say "approximately" because I will typically identify such losses by touch.

The exceptions to the above are hydrogen and $CO_2$. They have negative Joule–Thompson expansion coefficients. My only experience with such gases is with 99% pure hydrogen at the hydrogen plant in the Carib Island refinery. The 1000-psig hydrogen supply header leaked into a lower 850-psig pressure system. The hydrogen heated quite a bit upon expansion.

## FIXING LEAKING RELIEF VALVES

This book is about how we, as technical people, can operate and design process equipment in an environmentally responsible way. One simple concept is to install relief valves with isolating block valves. The relief valve sits on top of a full-size gate valve. The ASME (American Society of Mechanical Engineers) boiler code states that this is legal provided that the gate valve is "chain locked open." A chain is passed through the valve handle and the valve stem guide. The key to the lock is then retained in the shift supervisor's office. This is expensive, because we have then to install a spare relief valve so that the vessel relief valve can be repaired on-stream. Typically, we might have three half-sized relief valves.

Another way to reduce hydrocarbon losses through leaking relief valves is the use of rupture disks. A rupture disk is thin, flexible, and will fail as soon as the relief valve opens. Of course, it's a single-use item. So once the relief valve pops open, the rupture disk no longer will prevent leaks. Still, it is a useful method to retard hydrocarbon losses to the flare.

When I came to work as an operating superintendent at the alkylation unit in Texas City in 1974, I had a leaking relief valve on my unit. I saw it right off. It was hard not to see; the 6-in. line connecting the relief valve to the refinery flare header was iced up for 20 or 30 ft. The process line itself was alky feed. Alkylation unit feed is mixed butane and butenes at about 100 psig. I tried very hard not to see the leaking relief valve, but to no avail. It bothered me every day. I would have fixed it during my first week in Texas City, but unfortunately, there were no isolation gate valves under any of my relief valves. Certainly, I was not about to shut down the world's largest alkylation plant to fix that one leak.

Finally, after a month, I hit upon a comforting solution. The relief valve had been leaking when my predecessor, Paul Stelly, was the operating superintendent. Hence, it should have been repaired before Paul turned command of the unit over to me. And I could not worry about other people's failures and mistakes. But still, it bothered me for over a year, until I finally did fix the leaking relief valve during an alky unit shutdown.

Now, 35 years later, using the calculation technique summarized above in the section "Measuring Leaks Through Valves," I have estimated my rate of liquid butane leakage to be about 20 to 30 bsd $\pm$ 50%. That is not such a big leak in a process line flowing over 20,000 bsd. But between the time I noticed the problem and the time I fixed it, I probably flared 10,000 barrels of alky unit feed. In retrospect, it's strange that no one at the Texas City plant suggested that I had ought to do something to stop the flaring loss of butanes and butenes.

Today, though, I would never tolerate such waste. Here's what I would do. If I did not have an existing $\frac{3}{4}$-in. bleeder below the relief valve, I would have such a bleeder installed on-line. This requires drilling a hole through a packing gland that is clamped or welded to the outside of the process pipe. I have done this sort of thing at home recently when I installed the new water feed line to our expensive and useless ice maker.

Next, depending on the hydrocarbon service, start a flow of vapor into the new connection beneath the relief valve. For the defective butane relief valve in Texas City, I would have used nitrogen. I estimated that about 2000 scf/day of nitrogen would have continued to flow into the flare header. Alternatively, natural gas could have been used if it was available at over 100 psig. For heavier hydrocarbons, at a greater temperature, steam could be employed for the same purpose. We do this all the time on delayed coker coke drum relief valves.

One thing I do know for sure. I would not now wait over a year to fix such a problem or waste my time blaming it on Paul Stelly. The creator of the Universe, I am quite certain, would never accept that as a valid excuse for inaction.

## MEASURING FLOWS IN FLARE LINES

Refinery flare headers are usually fed by several branches. Knowing the approximate flow in each branch is necessary to track down the source of flaring losses. I developed the following simple procedure to do this at the Good Hope refinery in Louisiana. Equipment needed is a container of Freon or any other tracer gas such as lithium bromide or sulfur hexafluoride. Also, the appropriate tracer handheld detector is needed. Proceed as follows:

*Step 1.* At the start of the flare branch piping, admit the tracer gas for a few seconds.

*Step 2.* An engineer or technician located at the terminus of the line will monitor the line flow with his tracer gas analyzer.

*Step 3.* As soon as the tracer gas is released, the downstream observer is alerted by radio. He then notes the time for the peak flow of the tracer gas to pass his location.

*Step 4.* From the time and distance involved between the release of the tracer gas and its passing the downstream point, the flowing velocity can be calculated. Knowing the diameter of the pipe, the volume of gas flow can then also be calculated.

*Step 5.* Sample the gas to determine its molecular weight. Now the weight of gas flow through the branch piping of the flare system may also be calculated, correcting for the flowing temperature.

## VALVE STEM PACKING LEAKS

My shift operators in Texas City hated me. For example, I used to walk around my sulfuric acid alkylation unit with a crescent wrench, with which I stopped packing leaks on valve stems. The valve stem on a gate valve passes through a packing gland. That's just rings of rope stuffed down around the stem. On top of the last ring there is the packing gland follower, which compresses the rings down onto the stem. The follower is, in turn, compressed by two hex nuts. If the hydrocarbons are leaking around the stem, tighten up both hex nuts to stop the leak. But if you tighten up the hex nuts too much, it makes the valve harder to turn. Which is why my operators hated me.

"Look, guys," I would say, "If the valve is too hard to turn, back-off on the hex nuts. Then turn the valve. Afterward you can retighten the packing gland follower."

And they would say, "Lieberman, why don't you go back to New York City with the rest of them Damn Yankees? And take your [expletive deleted by publisher] crescent wrench with you."

## LEAKS INTO COOLING WATER

In Figure 18-3 I've shown how to detect leaks into cooling water when the return header pressure is less than atmospheric pressure. Ordinarily, cooling water flowing back to a cooling tower is under a positive pressure of 10 to 40 psig. In this case I would first check for hydrocarbon emissions with a portable gas test meter at the cooling tower return header outlet. If I find hydrocarbons, I'll go to the water outlet from each cooler and crack open the high-point bleeder valve, and check again for hydrocarbon emissions. Excessive rates of organic fouling in the cooling-water system, or excessive use of chlorine, are both indications of hydrocarbon contaminating leaks into the cooling-water circulating system.

## FLANGE LEAKS

I always resent flange leaks. It's an indication of sloppy workmanship. One problem that I've encountered is the use of paper-type flange gaskets. These gaskets are intended for blinding (spading) lines during turnarounds. That is, for temporary, low-temperature service. It's the job of the process supervisor to ensure that the correct gaskets are used on piping flanges, manways, and nozzles.

The reason for most flange leaks is something much simpler. That is, the flange surfaces are not cleaned properly. Or, the flange bolting was not tightened in the correct sequence. It's possible to install a clamp over a leaking flange, but that's quite expensive. I have had flange leaks in propane and isobutene service that I tolerated for months. One isobutane leak became progressively worse until the East Plant operating manager, Frank Citek, visited my unit one afternoon. "Lieberman," he screamed, "are you crazy? Shut this plant down. And I mean now!"

## AIR COOLER TUBE LEAKS

One common cause of tube leaks in air coolers is thermal stress. This can result in the tubes being partly pulled out of the header box tube sheet. The thermal stress is a consequence of the inlet and outlet process nozzles being on opposing sides of the air cooler. This causes the process piping, as it changes in temperature, to push and pull on the air cooler header boxes. But the piping itself constrains the free movement of the air cooler.

I've had this problem most recently on an amine regenerator overhead acid gas condenser (i.e., $H_2S$) in Lithuania and a delayed coker quench tower condenser in California. In both cases the solution was to use a tube bundle with an even number of tube-side passes. An air cooler with an odd number of tube passes will have the inlet and outlet nozzles on opposite sides of the bundle. With an even number of tube passes, one end on the tube bundle can be left free to slide across the back end of the air cooler support structure. This will greatly reduce the tendency for piping stresses to cause tube leaks and hydrocarbon emissions.

## LEAKING PUMP MECHANICAL SEALS

This is a huge subject to which I will make my own tiny contribution. I noticed at Texas City that my propane and butane pump mechanical seals often leaked. These streams were saturated with water. As the seals leaked, moisture would crystallize to ice between the seal faces. The tiny ice crystals would force the seal faces farther apart, which caused more leakage. As the light hydrocarbons flashed across the seal faces, they auto-refrigerated, which then caused the ice crystals to grow.

Then old Zip, my ancient foreman on D shift, would play a steam hose on the seal to melt the ice crystals and stop the seal from leaking. Which was okay if D shift was working. So my idea was to allow a whisper of steam, from a $\frac{1}{4}$-in. copper tubing, to keep the seals on the propane and butane pumps warm all the time. This prevented ice crystals from initiating the hydrocarbon leakage. But it was necessary to be careful not to blow the steam against the bearing housing, as that world contaminate the lube oil with water.

## FIXING WELD LEAKS

"Old Zip would say, "If you have enough drips, you could fill up an ocean." He taught me a method to stop hydrocarbon leaks. Once we had a small leak on a weld on a carbon steel pipe. It was a 6-in. line in 100-psig isobutane service. Zip used a plug of wood with a sharp point. He tapped it forcefully into the small hole. This stopped the isobutane leak. Next, he used an epoxy mixture and a roll of fiberglass cloth to wrap the pipe and the wood plug. Finally, Zip dried out the cloth impregnated with epoxy under a heat lamp for about 4 hours. This primitive repair lasted for many months, until the next unit shutdown when I could replace the defective piping spool piece. I also used this technique in Texas City to fix $H_2SO_4$ acid leaks using metal wood screws (see Chapter 21).

# Chapter *20*

## *Composition-Induced Flooding in Packed Towers: FCU Fractionator Expansion*

I had worn my new blue suit for my first day at work as technical service manager at the Good Hope refinery in Norco, Louisiana. My office was not much to look at. Still, it was a step up from my cubicle at Amoco Oil. As I unpacked my books, the phone rung.

"Lieberman, this is Jack."

Great. The owner of the refinery himself, Jack Smartly, was calling to welcome me to the plant. "Good morning, Mr. Smartly. I'm just settling in."

"What are you doing in your office? The cat LCO (fluid catalytic cracker light cycle oil) is black. Get down to the unit and figure out what's going on. And call me back in 20 minutes!"

"Okay, but I was just going to. . . ."

"And take off that suit!" Mr. Smartly shouted, as he slammed down the receiver.

I looked at my watch. It was 8:25 A.M. I had been the tech manager at Good Hope for 25 minutes. The owner of the refinery was already shouting at me. What would the future bring?

### FLOODING OF THE SLURRY OIL PUMP–AROUND SECTION

Figure 20-1 is a sketch of the FCU fractionator's lower section. The control room panel $\Delta P$ recorder indicated that the pressure drop across the slurry pump-around

*Process Engineering for a Small Planet: How to Reuse, Re-Purpose, and Retrofit Existing Process Equipment,* By Norman P. Lieberman
Copyright © 2010 John Wiley & Sons, Inc.

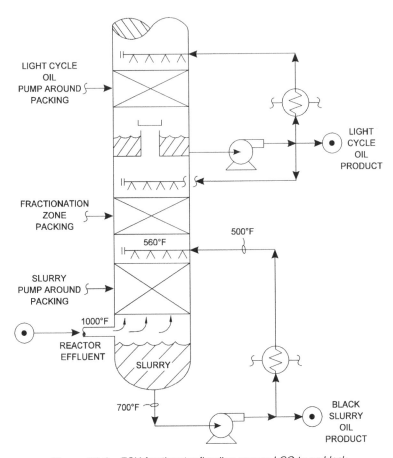

***Figure 20-1***   *FCU fractionator flooding causes LCO to go black.*

packed section was 32 in. $H_2O$. The height of the slurry pump-around packing was 8 ft. Thus, the slurry section packing $\Delta P$ was 4 in. of liquid per foot of packing. Above 2 in. of liquid per foot of packing, the type of packing used in the slurry section, would flood.

The $\Delta P$ value across the fractionation zone between the slurry oil and LCO pump-around sections was 3 in. of liquid per foot of packing. Thus, this fractionation section was also flooding. As both the slurry pump-around and fractionation packed zones were flooding, black slurry oil components would be entrained into the light cycle oil pump-around. Armed with these data, I walked into Jack Smartly's cramped office.

"Mr. Smartly," I began, "First allow me to say how much I appreciate working in the Good Hope. . . ."

"Get to the point, Lieberman," the owner yelled. "Is the LCO still black?"

"Yes it is. The slurry pump-around packed section is in fully developed flood, as indicated by the high-pressure drop. Sir, flooding progresses up the tower, so the slurry vs. LCO fractionation section is also flooding. This is causing black slurry oil components to be entrained up into the LCO pump-around section."

"So, Lieberman, is the LCO pump-around packed section also flooding?"

"No, it's not. The $\Delta P$ across the LCO pump-around packing is 0.5 in. of liquid per foot of packing, which is normal. But the entrained slurry oil is contaminating the yellow LCO with black slurry oil."

"But why is the slurry pump-around section flooding?" Mr. Smartly asked.

"At the 65,000-bsd feed rate to the FCU, the vapor rate inside the slurry oil pump-around section is too high, by about 10%. I can stop the flooding if you will permit me to reduce the unit feed to about 58,000 bsd."

"Lieberman, you're a real idiot," Mr. Smartly screamed. "Don't you know that we just revamped this fractionator to operate at 80,000 bsd of feed? Don't you realize that Glock Engineering designed the tower packing and the internals for the 80,000-bsd rate? Are you crazy, or stupid, or both? Don't tell me that we can only run 58,000 bsd after I spent millions of dollars revamping the fractionator to run at 80,000 bsd of feed!"

Somewhat fearfully I replied, "I'm sorry, Jack, but Glock Engineering did not do the process calculations correctly. To be specific, they failed to take into account the bubble effect."

"The what? The bubble effect! What is the bubble effect?"

"The problem, Jack, is composition-induced flooding. I'll try to give you a simplified explanation by referring to Figure 20-1.

*Step 1.* The reactor effluent is 100% superheated vapor at 1000°F. This vapor needs to be cooled to 700°F to condense the slurry oil product, which is only 2% of the reactor effluent vapor.

*Step 2.* The slurry oil pump-around packed bed is 8 ft high. Thus, it contains about three theoretical fractionation stages. As the 1000°F vapors mix with the slurry oil pump-around liquid on the lowest stage, the slurry oil will partially evaporate.

*Step 3.* The lightest portion of the slurry oil pump-around is actually LCO. So it's LCO vapors that are generated from the evaporating slurry oil. The 1000°F vapors cool off to about 700°F by evaporating the LCO in the slurry pump-around.

*Step 4.* To cool off a pound of the vapors from 1000°F to 700°F releases about 200 Btu, as follows:

$$(1000°F - 700°F) [0.67 \text{ Btu/(lb-°F)}] = 200 \text{ Btu/lb}$$

*Step 5.* The latent heat of evaporation of the LCO is about 100 Btu/lb. Thus, each single pound of 1000°F reactor effluent vapor creates two additional pounds of vapor inside the slurry oil pump-around packing.

*Step 6.* The moles of LCO vapor add on to the moles of reactor effluent vapor and the total vapor flow is cooled to 550°F by the 500°F slurry pump-around return liquid. As the vapors are cooled in the top of the slurry pump-around packing, the 2 lb of LCO vapor recondenses.

*Step 7.* The condensed LCO flows back down the packing to be revaporized by the superheated reactor effluent 1000°F vapor. The LCO vapor is trapped inside the slurry oil pump-around packing between the superheated 1000°F vapor at the bottom and the 500°F pump-around return liquid at the top."

"Mr. Smartly, the bubble of vapor generated is due to the more volatile, lighter material in the slurry oil pump-around. It's this bubble of vapor that causes the packing to flood. That's why I call it composition-induced flooding. When Glock Engineering calculated the vapor loads for the bottom of the fractionator they used 800,000 lb/hr vapor at 1000°F. But they should have used 2,400,000 lb/hr vapor at 700°F. And that's why the fractionator floods at less than 80,000 bsd," I concluded.

Jack Smartly, the owner of the Good Hope refinery, stared at me with naked hatred in his eyes. "You [expletive deleted by publisher] engineers. Now I'll have to build a new tower. I'll make it twice as big. Then I won't have to worry about your bubble effect. I should have all you engineers shot."

"Mr. Smartly! Don't blame me. It's inherent in the process. It won't help to shoot me. It wouldn't happen if the slurry oil was a narrow boiling-range mixture. But it's not necessary to build a new tower. I have a solution to the flooding using the existing tower."

## FRACTIONATOR VAPOR LINE QUENCH

"Okay, Lieberman. Let's hear your idea. We can use it until the new 24-ft-diameter tower is built. That will probably take six months."

My idea is summarized in Figure 20-2. It's the external vapor line slurry oil quench. About half of the 500°F slurry oil pump-around return is pumped into a perforated pipe inserted in the vapor line. The holes in the perforated pipe all face the direction of flow to promote better vapor–liquid mixing. By "de-superheating" the vapor from 1000°F to 780°F upstream of the column, the bubble effect is almost eliminated. Thus, the maximum vapor flow inside the slurry oil pump-around packing will be cut almost in half.

"Look, you dumb engineer. That sounds like a good idea," observed Jack Smartly, "But won't the liquid in the vapor line coke-off? We keep the vapor line very well insulated now to prevent small amounts of liquid condensation, as the condensed liquid will turn to coke."

"Jack," I replied, "You're right. But this is a large amount of liquid, almost equal to the weight flow of the vapor itself. Also, as long as the vapor is quickly quenched down below 800°F, rates of coke formation due to thermal cracking will be quite low."

"How can you be so sure?"

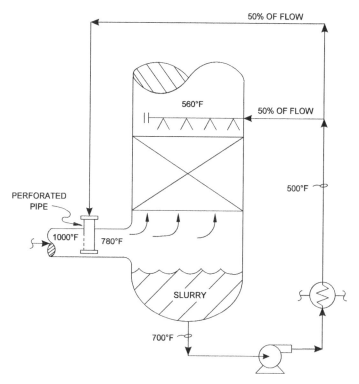

**Figure 20-2** *External quench reducers vapor load to packed bed by de-superheating vapor.*

"Because, it's the same idea that we use in delayed cokers. We quench the coke drum overhead vapors using cold slop or gas oil, from 820°F to 780°F, to suppress coke formation in the coke drum overhead vapor lines."

Mr. Smartly looked quietly at me. He rearranged the colored pencils scattered on his desk to correspond to the colors of the rainbow. "Well, Lieberman, you thought up that vapor line quench just now? Just sitting here? You must be pretty smart. I guess I'd better not have you shot after all."

"Mr. Smartly, perhaps you should know something about my background. I have been inspired by Zeus. I am an agent of the divine will on Earth!"

Actually, I had been inspired more by Ashland Oil than Zeus. I had read in the NPRA Q&A (National Petroleum Refiners Association question and answer session) meeting summary that Ashland's refinery in Cattlesburg, Kentucky, had installed such a vapor line quench on their cat cracker—and that it had worked. But perhaps Zeus had inspired me to read that particular section of the NPRA Q&A (published every year for many decades and now available on the Internet).

"Jack, we can install this new quench nozzle on-stream. Let's hot tap a new 12-in. gate valve on top of the vapor line. Then we can insert the new quench pipe into the vapor line through a packing gland. All the external piping will just be 12-in. carbon steel pipe. I'm sure that this will allow us to run 80,000 bsd of FCU feed, which is sufficient for our plant capacity."

"Well, Norm, as you've been inspired by Zeus, who am I to oppose the Immortals' will? Just be careful with the hot tap. That's a high-chrome-steel line. Let's make sure that the welding is all done to conform to the ASME code. Also, don't forget that the 1000°F vapors, if they leak, are above their autoignition temperature."

"So, Jack, can we forget about the new 24-ft- diameter tower?" I asked.

"No, Norm! Faith has its limits. I have faith in divine power, but somewhat less in you. We'll cut a few corners and have the new tower up and running in six months. We'll try your idea also. Norm," Mr. Smartly concluded, "one must always hope for the best, but plan for the worst."

## HOT PUMP PIPING STRESS ANALYSIS

We borrowed a large hot tap machine from Shell's Norco Refinery across the road. The new vapor line quench worked just fine. I controlled the 500°F slurry quench flow to hold the vapor inlet temperature to the fractionator below 780°F (see Figure 20-2). Feed rates of 80,000 bsd were achieved while producing a clean, pale yellow, LCO product and without flooding in the slurry oil pump-around packed section.

Eight months later, the new giant fractionator was ready for service. It towered into the sky over Norco, Louisiana—rather like a Cajun Tower of Babel, an offense to Zeus, and a monument to human folly.

Mr. Smartly had indeed cut a few mechanical engineering corners to erect the tower so quickly. One of these was designing the suction piping on the three slurry oil circulation pumps without adequate thermal flexibility (i.e., without any of the normal piping stress analysis). Heating steel pipe from 70°F to 700°F creates thermal stress at the elbows. The more complex the piping, the more complex the engineering calculations required to design the suction piping to avoid piping stress failure. This is especially true for heavy wall, large-diameter chrome piping. So when the fractionator was started up with all three slurry pumps in service, all the suction piping heated and expanded at a uniform rate and all was well. But when one of the three pumps was shut down a week later for maintenance, the suction piping of an adjacent pump failed due to excessive thermal stress. The 700°F slurry oil autoignited. The resulting fire destroyed this modern tower of Babel.

I still recall seeing the black smoke rising across the swamp, while driving west on I-10. I remember telling my son, "See, Joe, Father Zeus's will shall be done on Earth, as it is in heaven."

"Dad, are you going to start in with that dumb Mount Olympus stuff again?" asked Joe.

"Yes, my son," I answered. "I am on a mission from the Earth Mother. I truly set Jack Smartly on the path to righteousness with my vapor line quench. It worked and fulfilled his need to run 80,000 bsd of cat feed. But the Devil filled his heart with lust for 90,000 bsd. So in his pride and arrogance, he built a new tower that rose into the heavens. And now the Immortals have cast down the tower into the dust and humbled Jack Smartly. Joseph, pride goeth before the fall."

"Dad, I guess Mr. Smartly should have been satisfied with the 80,000 barrels, using the tower that he had. I know Grandma Mary would never have given him any money for the 90,000 bsd new tower. Grandma is too cheap."

We pulled over to watch the dense black cloud creep slowly up the Mississippi River. "Dad," Joe asked, "are you really working for the Earth Mother and Zeus on this environmental stuff, or have you really gone totally crazy?"

"Son, I'm working as a subcontractor on a big project, the scope and purpose of which we can never fully comprehend."

## SHALL WE GAMBLE ON THE FUTURE?

Jack Smartly and I became friends. I worked for him for eight years. He promoted me to refinery manager and then vice president. I worked with his charming wife, erratic children, and confused brother. Yet the conflict continued. I, the pessimist, wanted to make small, incremental improvements to existing process equipment and to existing natural gas production wells. My idea of progress was to add a timer (emitter) to a natural gas well-head so as to blow condensate and brine out of a well bore automatically, twice a day [1].

Jack's idea of progress was to drill a 23,000-ft gas well to tap into an imaginary reservoir; or, to force the world's biggest frac job (i.e., injecting sand into a tight rock formation to promote gas flow) in Zapata, Texas.

Finally, I began to understand that our conflict was not technical in nature, but a reflection of our different world views. My family came from the narrow, confined ghettos of Eastern Europe. I saw our planet as a small place with limited resources, and life in a precarious balance. Jack's family came from the open glens and vales of highland Scotland. He saw unlimited opportunities in every circumstance. The bigger the bet, the bigger the payoff: the gambler's creed.

Jack called me a pessimist, and would often eject me from critical planning meetings. I called him a wild-eyed optimist, who refused to consider the inevitable difficulties that arise in any project. And in the larger frame of reference, isn't this the same conflict we have today on our little planet? Me, I never gamble.

## PERCEPTION VS. REALITY

Process engineering is different than other sorts of applied technology in that detailed calculations are often an exercise in futility. It is almost always better to ratio or interpolate from field data rather then to calculate expected results using engineering equations. Let me provide some examples:

- *Piping $\Delta P$.* If you ever calculate a pressure drop through a pipe, you will need to provide the friction factor. To arrive at the appropriate friction factor, you must first know the interior surface roughness of the pipe. I doubt that you have crawled into a 6-in. pipe to measure its interior surface roughness. You just have to guess at the surface roughness, which means that you are, to some extent, guessing at the $\Delta P$.

- *Heat exchanger surface area.* It's true that the heat transfer surface area of an exchanger is to a large extent a function of the heat transfer coefficient. But in most industrial services, the heat transfer coefficient is mainly a function of the fouling resistance. And how does one know the actual fouling resistance? Of course, we don't. So when we calculate a heat exchanger size, we are really just guessing at the required surface area.

- *Distillation tower: number of trays.* You would think that the number of theoretical contacting stages assumed in our computer model calculations would permit us to determine the number of trays required in a new distillation tower. Not true! To convert from the number of theoretical contacting stages to the number of actual trays, you have to divide by the tray efficiency. The tray efficiency used is only a guess. Thus, you're really just guessing at the required number of trays in the new tower you have just designed.

- *Distillation tower diameter, reflux rate, and reboiler duty.* A huge software industry has developed to allow the process engineer to model fractionation processes to size the heat duties of the overhead condenser, reboiler, and tower diameter, with precision. However, if you have (as most of us must) run such models, you will have had to input into the model an equation of state. The equation of state is used to calculate the relative volatility. The relative volatility is used, in turn, to calculate the required reflux rate, to produce the required degree of fractionation. The reflux rate then determines the tower heat balance (i.e., the condenser and reboiler duties) and diameter. But how did you determine the equation of state? Well, you guessed at it. Thus, you have guessed at the entire design.

Composition-induced flooding adds another layer of confusion, in addition to tray efficiency and relative volatility, when applying distillation theory to actual design problems. Tray vapor and liquid loads are radically affected by wide variations in latent heat, temperature, and molecular weights of a wide range of components. Process engineering calculations are rendered progressively less meaningful by the complexity of the problem. Eventually, our ability to grasp the fundamental nature of the problem is drowned in this vast sea of complexity. The real vapor–liquid rates are unknown and unknowable.

All our process engineering is like that. It's all guesswork. Often, it seems as if we have calculated our engineering parameters with great precision because we produce lots of numbers on a computer screen. But it's all a grand illusion.

Far better to spend our limited time in the field, understanding the process as it exists in reality rather than optimizing a false perception of reality that exists only in the digital world. If the Creator ever asks me what the fundamental error that process engineers have made on our little planet is, I'll say, "Lord of the Universe, they have designed new process equipment without reference to the operation of existing equipment in a similar service. They have sinfully used computer simulations in their comfortable, air-conditioned offices rather than interacting with hostile plant operators, dirty process equipment, and foul smelling fumes. They have used their university degrees as a shelter from the harsh winds of reality."

## REFERENCES

1. Lieberman, N. P. *Troubleshooting Natural Gas Processing: Wellhead to Transmission*, PennWell Publications, Tulsa, OK, 1987. Reprinted by Lieberman Books, 2008.
2. Lieberman, N. P. "Use of Structured Packing in FCU Fractionator," *Hydrocarbon Processing*, Apr. 1984.

Chapter 21

# Maintenance for Longer Run Lengths

Ms. Sandy Bellville
Materials Supply Manager
Hert Chemicals Plant
Houston Ship Canal
Houston, Texas

Dear Sandy,

I hope you will remember that I will always treasure our relationship of the past two years. My feelings for you and Hert Chemicals run deep. But for the last six months, you have probably sensed that I've become distant. I suddenly remembered this morning that I haven't phoned you since Christmas to order any fresh sulfuric acid, or even to return any spent acid to you for regeneration. It's not that I don't care any more. We're still the same people.

I haven't found someone new. Amoco and Hert Chemicals have an exclusive contract for the regeneration of spent alkylation and jet fuel treating $H_2SO_4$. It's just that I've been able to get my acid regen plant working on a consistent basis. I suppose you were happier when you were with Paul Stelly, the previous operating manager of the acid regen plant. Paul was a fine manager. I recall that Paul used to faithfully order a barge of acid every week, and I haven't in five months. I'm so sorry, but let me explain.

*Process Engineering for a Small Planet: How to Reuse, Re-Purpose, and Retrofit Existing Process Equipment.* By Norman P. Lieberman
Copyright © 2010 John Wiley & Sons, Inc.

## ACID PLANT OPERATING FACTOR

For years, Amoco Oil's $H_2SO_4$ regeneration plant in Texas City had been limping along with an on-stream factor of 30%. I became the new manager on April 1, 1974. My capital improvement budget for the plant shown in Figure 21-1 was zero. Most of my operating budget was used to pay for regenerating excess spent sulfuric acid in the Hert Chemicals plant in Houston. Once or twice a week, Sandy Bellville would have a barge of fresh acid delivered to my tanks on Snake Island in Galveston Bay. The barge would sail away with a load of spent alky acid to the Houston ship canal.

I can't say exactly why I hated to call Sandy for the acid barge. It was a humiliating experience to admit that I couldn't regenerate my own spent acid. It wasn't the expense. Amoco had lots of money. It was, rather, a matter of a failure to prove my manhood. Kind of a macho-type thing.

After a year of folly, frustration, and failure, I woke one morning inspired by a clear and forceful plan. I would investigate each outage in detail, and devise a plan to prevent its reoccurrence. I would not rely on my shift supervisors or maintenance foreman to handle problems. I was not going to waste any more time writing reports, preparing budgets, or attending meetings. I'd been inspired by a dream I had that night. In my dream, my mom was talking to Mrs. Harris, Gloria's mother.

"Silvia, you should have Gloria meet my son Norman. She's such a lovely girl."

"So! That skinny boy's your son? Norman?" said Mrs. Harris.

"Gloria's a real beauty and Norman's a real genius. He can fix anything. One new tube. Presto! The TV's perfect!"

## LEWIS PUMP WEAR RINGS

The single biggest problem that I had on my plant was low acid circulation rates in both my drying tower and absorber (Figure 21-1). The low circulation rates left residual $SO_3$ (sulfur trioxide) in the stack plume. The $SO_3$ reacted with atmospheric moisture to form a heavy, white plume of $H_2SO_4$. The plant manager, Mr. Dorland, would see the plume and call down to my unit.

"Either stop that plume or shut down immediately," Mr. Dorland would order.

Then, I would pull both Lewis circulating acid pumps and send them to the maintenance shop that specialized in the nonrepair of rotating equipment. The pumps always ran fine when they were placed back into service to circulate 93% and 99% $H_2SO_4$. But their circulation rates fell dramatically within weeks. So I decided to visit the pump repair shop to determine what was wrong.

"Mr. Norm," Leroy the machinist explained, "It's them there impeller wear rings again."

"What's an impeller wear ring, Leroy?"

"Well," Leroy answered, "It's a ring that fits around the eye of them pump's impeller. It reduces internal recirculation of liquid between the impeller vane tip and the pump suction."

"What's wrong with my Lewis pump's wear rings, Leroy?"

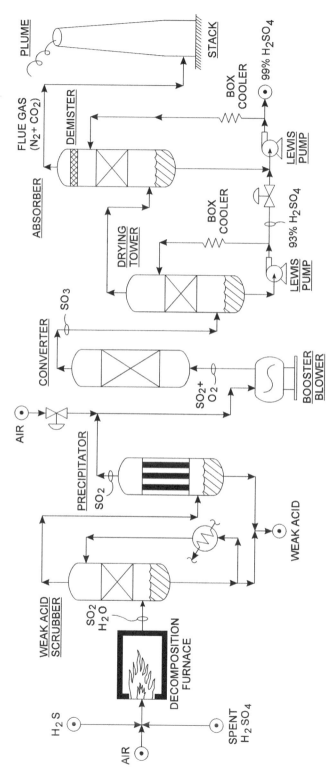

*Figure 21-1* Spent sulfuric acid regeneration plant.

"Mr. Norm, they eat-up real fast. That there acid is real hard on them wear rings."

"Okay, Leroy. But what are the rings made of? Some kind of steel?"

"No sir! We make our very own wear rings. We make's them out of Carpentor 20 metal," Leroy said proudly. "Look here, Mr. Norm. Them wear rings that there Lewis Company provides are no good. Just look how they just breaks-up." Leroy demonstrated how an Illirum wear ring from the pump manufacturer would shatter on the shop's cement floor when dropped 3 ft. The Carpentor 20 ring, on the other hand, bounced happily intact off the cement.

So I called the Lewis company. They told me that while Illirum metal was extremely brittle, it was also extremely resistant to strong acid attack at high velocity. They suggested that I not bounce their wear rings off cement floors, but install them in their pumps. Using Carpentor 20 in high-velocity, strong-$H_2SO_4$ service, for impeller wear rings, was certain to result in high rates of erosion, and thus increased rates of internal recirculation and loss of pumping capacity.

The maintenance people had started to make their own pump parts at Texas City a few years ago to save money. They had considered that Carpentor 20 metal was a suitable material for strong $H_2SO_4$. But this was not the case for a Lewis pump impeller wear ring. Using the manufacturer's Illirum rings eliminated one of our major shutdown problems. I learned then that it's best to buy replacement pump parts from the original pump manufacturer.

## PRECIPITATOR STAR WIRE FAILURES

The electrical precipitator shown in Figure 21-1 was used to knock-out entrained, low-strength acid mist droplets from the feed gas flowing to the vanadium pentoxide extruded catalyst converter. Strong $H_2SO_4$ could not be produced if the precipitator was not working. It consisted of several hundred 12-in. lead tubes acting as the anode. The cathode was a steel-cored lead wire strung down the center of the lead tube. A wire would break once every few weeks. The broken wire would then arc and short-out the precipitator. The plant would be shut down. By the time a maintenance guy could be called out from home to cut out the broken wire and plug the tube, four to eight hours would elapse. By this time, the converter would have cooled off below the reaction temperature needed to produce sulfur trioxide:

$$SO_2 + \tfrac{1}{2}O_2 = SO_3$$

This reaction is exothermic. The heat of reaction would keep the converter hot enough to sustain the reaction. But the converter would cool below reaction temperatures after four idle hours. Then, the startup heater would have to be lit, and the long, complex startup procedure, taking two days, would begin again.

I solved this problem by direct action. I picked up my wrench and hacksaw at home. I violated every Amoco safety regulation and unbolted the precipitator manway. I cut out the broken lead wire and plugged the tube with a wooden plug. Within two hours of being shut down, the plant was up and running again. I still keep a piece of the old lead wire in my desk.

## CONVERTER INSULATION

The converter vessel was so badly insulated that it would drop below its reaction temperature during a heavy rain. Also, each shutdown of more than four hours would also cool the converter below its reaction temperature. Either way, two days of production would be lost while we rebooted the unit. I had the vessel insulation and weatherproof metal sheathing repaired. The insulating outfit nicely waterproofed all gaps with roofing tar. The $SO_2$ converter now ignored wind and rain. But of greater importance, we could now experience shutdowns of eight hours before the $SO_2$ converter vessel cooled below its reaction temperatures. Also, the converter looked much more professional.

## OPERATOR PSYCHOLOGY

The incident with the broken precipitator wire and the converter's new silver and black weatherproofing changed the operators' attitude toward the regen plant. Instead of looking for excuses to shut the plant down, they began to look for ways to keep it running. For example, Joe Hensley, the unit inspector, discovered that the demister was loose. As shown in Figure 21-1, the demister was at the top of the absorber. The demister never seemed able to suppress the acid plume from the stack. Every shutdown I would inspect it myself, but it looked fine. But during one brief outage, Joe Hensley climbed into the absorber and noted that one side of the demister was loose. The gas flow would push the unfastened side open during operations. But when the unit was shut down, the demister would fall back into place. A few bolts eliminated this big operating problem. Thanks to Joe, the number of angry phone calls from Mr. Dorland greatly diminished. I learned to respect the efficiency of demisters in nonfouling and noncorrosive services. Incidentally, strong sulfuric acid is far less corrosive than weak sulfuric acid.

## FIXING AND PREVENTING PROCESS PIPING LEAKS

There are two ways to fix a pinhole leak in a cast iron section of acid piping:

- The Amoco Oil way
- My way

The Amoco Oil method, in widespread use in the refining industry and at Texas City to fix a leak, was to shut down and replace the leaking piping section with a new section. My method was:

- Get Kenny Trahan, the maintenance foreman, to put on an acid-proof suit.
- Using a stainless steel wood screw (we used a brass screw for hydrocarbon leaks), screw it into the leak until the flow stopped.

- Cover the screw and adjacent piping with epoxy cloth and resin. Apply a heat lamp for a few hours.
- Make up a replacement piping spool piece and install it the next time the plant shuts down for another reason.

I became famous in Texas City for fixing piping leaks. I was even called the Epoxy King by my peers. I noticed that half my leaks were on cast iron elbows and fittings. The leaks were almost always at a small mark left on either side of the elbow associated with the casting process. During the next unit outage, I had a $\frac{3}{4}$-in. threaded hole cut into all these marks on the process piping. An alloy steel plug was then screwed into the threaded holes, as were all the standby replacement cast iron piping, as a preventive maintenance step.

My policy regarding acid leaks as communicated to my shift foremen was:

- Try to get the shift operators to fix the leak.
- Next, call out a pipe fitter to fix the leak (typically, a waste of time).
- Call me at home, but do not shut down for an acid leak.
- If you have to shut down, first have a plan as to how to get the leak fixed within four hours before the converter vessel can cool off.

## USING WHAT'S AT HAND

The air intake at the suction of the booster blower shown in Figure 21-1 was not really needed. Upstream of this point, the sulfuric acid regeneration plant was under a vacuum. Combustion air drawn into the decomposition furnace provided the stoichiometric amount of oxygen needed to convert $SO_2$ to $SO_3$. The problem was that the air intake valve leaked. The air drawn in reduced the converter efficiency and temperature and made the stack acid plume worse. I climbed up to the air intake duct one day with John Brundrett, the maintenance superintendent.

"Lieberman, let's shut down and change the valve out. Let's do the job right," John screamed above the roar of the air leak.

I grabbed a bunch of rags out of my coverall pockets and let them be sucked up into the leaking valve. The roaring air was stilled.

"Lieberman, how you going to adjust your air flow now? Those rags are stuck in the valve."

"John, I'll just add or subtract rags as needed to maintain my stoichiometric ratio of $O_2$ to $SO_2$ to produce $SO_3$."

"Lieberman, did anyone ever tell you that you're crazy?"

"Well, John, not today. You're the first."

The problem was that I reported my rag innovation in my official monthly progress report to headquarters in Chicago. I can still hear Mr. Dorland's angry comment. "Lieberman! Did anyone ever tell you that you're crazy?"

## PRESERVING PUMP MECHANICAL SEALS

One of the really good things I did as an operating supervisor was to visit an acid plant run by a really experienced manager. One of the many things I learned was a simple trick to preserve the integrity of smaller centrifugal pump mechanical seals. Piping stresses sometimes pulled my acid pumps out of alignment. To avoid this problem, I learned to mount my pumps on rollers. The rollers were just a few short sections of 3-in. pipe. I also learned the value of having a knowledgeable, hands-on colleague to call on for practical advice. Although Amoco Oil operated 13 domestic refineries, the Texas City $H_2SO_4$ plant was the first (and the last) that Amoco built.

I mentioned before that some of my spent acid feedstock was used to treat jet fuel. In 1976, hydrotreating jet fuel was not common. $H_2SO_4$ was used to improve jet fuel color stability after the jet fuel had been caustic-washed to remove naphthenic acids. Some of the sodium would be carried over by the jet fuel into the spent acid. The sodium salts would plug the front-end tube-sheet of my decomposition furnace shown in Figure 21-1. Paul Stelly used to shut the plant down periodically to clean the plugged tube inlets. But I hit on a better plan. I segregated the small amount of sodium-rich spent acid from the jet fuel treater in a tank on Snake Island. When I had accumulated 1500 tons, I loaded it on a barge which sailed away across Galveston Bay to Sandy Bellville and Hert Chemicals.

I also figured out a way to dispose of excessive weak acid produced in the scrubber and precipitator. Paul Steely blended the weak acid back into his $H_2SO_4$ product, which weakened the acid to only 97% $H_2SO_4$. I once tried to blend the excessive weak acid production into Galveston Bay, but was caught in this environmentally hostile act after killing the bugs in the effluent-activated sludge treatment plant. I then hit upon a more progressive solution.

Strong (98 to 99%) $H_2SO_4$ was used to control the cooling-water pH in Texas City. I supplied the acid to the entire refinery. Rather than using good-quality fresh acid, I started supplying low-quality weak acid to control cooling tower pH, which worked just as well as the 98% sulfuric acid.

## H₂S AS SUPPLEMENTARY FEEDSTOCK

Texas City in 1976 did not have a sulfur recovery plant. All the hydrogen sulfide gas was flared—not too good a practice on a small planet. The $H_2S$ was supposed to be a secondary feedstock for the acid plant. The main feedstock was spent sulfuric acid from my 26,000-bsd alkylation plant. None of the previous supervisors had ever tried to process $H_2S$ on a continuous basis. However, it seemed to me that using $H_2S$ would:

- Make my decomposition furnace run hotter and thus decompose the spent acid more completely, to sulfur dioxide, which would reduce weak acid production.
- Reduce the relative amount of moisture in the furnace effluent compared to combusting spent acid, which was 10% $H_2O$.

- Increase acid production. This would offset acid losses that would otherwise have to be purchased from Hert Chemicals.

My efforts to consume the $H_2S$ that had previously been flared proved to be quite effective. For example, one day I phoned Hert Chemicals in Houston.

"Hi, Sandy, it's Norm."

"Well, Mr. Lieberman, what can I do for you today?"

"Sandy, you sound kind of angry."

"Norm! After everything! After all the acid we've exchanged. You haven't called in months. And then you sent that awful barge of contaminated, sludgy, sodium-rich, 65% acid to me. I'm very hurt."

"Gee, Sandy. It's in the contract. It's spent refinery acid. The contract states that. . . ."

"Norm," Sandy said in a softer tone, "Let's forget about that nasty barge. Our relationship can stand a single bad shipment. When do you want your next barge of fresh 98% acid? How about on Friday?"

"Well that's not exactly why I called. You see, Sandy, I've been running a bunch of $H_2S$ feed in my regeneration plant. Why flare it? It's bad for the planet. So, I kinda have lots of excess fresh 99% $H_2SO_4$. My 'cup runneth over' with sulfuric acid. My tanks are full up with fresh acid and I thought. . . ."

"What?" Sandy screamed, "You rotten, self-centered, uncaring S.O.B. You want to ship Hert Chemicals fresh acid? Did anyone ever tell you that you're crazy, cruel, and inconsiderate?!"

"Actually, yes, Sandy. It's been mentioned before. But as to my shipping excess fresh acid back to Hert Chemicals, it's in the contract."

But I think Sandy had slammed the receiver down before I could finish my sentence. Sadly, our paths have never crossed again.

## SUMMARY

Bhudda's last words before he died were, "Rely upon yourself." That's the lesson I learned in Texas City. Before I arrived, there were a variety of projects under study to replace the precipitator, the acid circulation pump, and the absorber. These projects all proved to be unnecessary.

Environmentally, I did pretty well, too. I eliminated 30 mt/day of $H_2S$ that was being flared, as the refinery did not have a sulfur recovery plant in 1976. The constant acid barge traffic across Galveston Bay between Amoco and Hert Chemicals slowed to a trickle and then stopped, which saved all the fuel for the barges. And although Amoco never rewarded me for my success, the experience has made me a more confident and self-reliant process engineer.

However, the lack of barge activity on Snake Island eventually created a problem.

"Mr. Norm, this here's Captain Mack callin."

"Okay, Captain. What's the problem?"

**Figure 21-2** Author is young man in black glasses third from left.

"There's a snake a layin on the top step of the pumphouse, down here by the acid dock."

"A what?"

"A rattler. Bout as big round as my arm."

"Well, leave it be. We're not transferring acid anyway."

"Mr. Norm. I can't rightly do that. My girlfriend's still inside the pumphouse."

And just in case you're thinking I'm making this all up—not about Captain Mack but about the acid plant—I've attached a photo of our March 1975 record-setting party. I'm the third guy from the left (Figure 21-2).

# Chapter 22

# Instrument Malfunctions

Too often, simple instrument malfunctions have precipitated major projects. A number of process design contracts that I have been awarded evaporated like the morning mist when exposed to the light of reality. Foremost among these lost and forlorn projects are problems initiated by misplaced liquid-level taps, lost instrument air signals, and malfunctioning control valves. I'll cite a few examples out of the hundreds that I have stumbled across in the last 45 years as a process engineer.

## CONTROL VALVE FAILURES

"Norman, will you be coming to our prayer meeting tonight?"

"No, Gloria. I don't think so. I don't actually believe too much in prayer."

"Norman, I'm surprised! I thought you're on a mission from Pallas Athena, Goddess of Wisdom. Please do come! We're meeting at 8 at my home."

"It's true I'm on a mission from Athena, but I've already got my instructions, so I don't need to report back until I've completed my mission," I answered.

Gloria turned off her computer and looked at the wall clock. "Well, it's time to go. We're having the gas oil recovery meeting in Tom's office this week. You need to come, too. Let's go."

"What's the problem?" I asked.

"Norman, you know the problem! There's too much light gas oil in the crude tower bottoms. The light gas oil rate is only 1500 bsd and it's supposed to be

*Process Engineering for a Small Planet: How to Reuse, Re-Purpose, and Retrofit Existing Process Equipment,* By Norman P. Lieberman

4000 bsd. I emailed you my crude tower simulation results. Didn't you open my email?"

"Sure. It was very definitive," I lied. "Having to revaporize all that light gas oil out of the crude tower bottoms is wasting lots of energy and vacuum tower capacity."

"Right," said Gloria. "That's what we'll be discussing with Tom. Dr. Beavorbrook wants us to shut down the tower in August and re-tray the gas oil section so as to maximize gas oil and distillate recovery."

"August. That's just two months away."

"Norman, you will be designing the new 317 stainless steel total trapout chimney tray. I'll provide the design for the expanded draw-off piping and product pumps. Tom will provide the specifications for the new fractionation trays above the flash zone, up to the diesel draw-off tray." Gloria glanced at her watch.

"Did I tell you that Dr. Beavorbrook himself may drop by our evening prayer meeting? Sure you can't make it?"

"No, Gloria, sorry. Also, I can't come to Tom's meeting either. I have to check something in the plant first."

## LOOSE INSTRUMENT AIR CONNECTION

I had noted that the gas oil draw-off control valve (Figure 22-1) was 100% open on the panel. But I knew that the valve position shown on the panel is not the real control valve position. The valve position displayed on the control panel is just an air signal indicating what the valve set-point position is supposed to be, not the actual valve position. So I climbed up to the fourth landing of the tower structure to check the valve position in the field. The light gas oil control valve was barely open. As shown in Figure 22-1, the valve was APO (air pressure to open), meaning that the instrument air supply tubing was connected underneath the diaphragm. Supplying

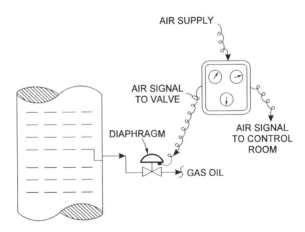

***Figure 22-1*** *Leaking air signal prevents product valve from opening.*

more air pressure to the diaphragm should cause the valve to open. The maximum air pressure was already flowing to the diaphragm. However, the little screwed tubing connection on the underside of the diaphragm had come loose. I suppose that I should have reported this to the instrument department for repair. But I simply tightened up the loose fitting with my tiny crescent wrench. The air pressure to the underside of the diaphragm increased and pushed the gas oil control valve wide open. Gas oil flowed rapidly out of the tower at a rate of over 5000 bsd.

Since I'm pretty smart, I figured I had better not tell anyone that I tightened up an instrument air connection without first getting approval from the unit operating personnel or the area supervisor.

## THE NEXT DAY

"Norm!" Gloria burst into my cubicle. "Great news. It's a miracle!"

"What's up?"

"Gas oil production doubled last night."

"Really, Gloria? How did that happen?" I asked.

Gloria's pale, thin face glowed with excitement. "It's all a testimony to the power of prayer. Dr. Beavorbrook will be so pleased. I spoke about our problem with reduced light gas oil production at last night's prayer meeting, which unfortunately you failed to attend. And now look! Our gas oil production has more than doubled in just one night.

"Spoke about it to who—God or Dr. Beavorbrook?" I asked.

Gloria was annoyed. "Norman, don't you believe in the Divine presence here on Earth?"

"Sure I do. I'm on a mission from the Goddess of Wisdom, Pallas Athena, myself."

## STUCK CONTROL VALVE

In 1968, a delayed coker fractionator I had designed started-up in Texas City. The distillate side draw-off product (Figure 22-1) was feed to a diesel oil desulferizer. The draw-off rate was designed at 3000 bsd, but the actual rate was only 1400 bsd. So I asked the panel board operator to open the draw-off flow control valve a bit more. He complied and opened the valve on the panel to 100%, but the flow remained at 1400 bsd.

Lab distillation results showed that the missing 1600 bsd of potential diesel oil product was dropping down my fractionator into the heavy coker gas oil product. Even in 1968, I knew that this was bad. The lighter distillate would be run through the fluid cracking unit, where 20% of it would be converted to catalytic coke and cracked gas. The coke would be burned off in the regenerator, and the cracked gas would be flared along with the rest of the excess refinery fuel gas. But nothing could be done, as I had opened the draw-off valve to 100%.

Ten years passed. I was visiting the Texas City refinery in 1978 to attend an important meeting, with many important people, to discuss important business matters. During the important two-hour lunch break, I walked over to visit my delayed coker fractionator. The same old panel board operator showed me that nothing had changed. The side distillate draw-off control valve was still 100% open, and the distillate flow was still only 1400 bsd.

However, now I knew that the valve position shown on the panel was not necessarily the real control valve position (unless there is a valve "positioner"). So I climbed up into the fractionator structure to look at the control valve. It was one-quarter open. The valve was an air pressure-to-open valve, as shown in Figure 22-1. There was a full 20-psig instrument air pressure to the valve, but the valve stem was stuck. Using a hand jack (a mechanical device used to move control valves manually), I opened the control valve to 100% and sprayed the stem with WD-40 lubricant.

When I went back to the control room, I saw that the flow of distillate had jumped to 2700 bsd. Another miracle! And don't think I've exaggerated about the 10 years. It really happened exactly as I've described.

## REDUCING CRACKED GAS LOADS TO VACUUM SYSTEMS

"Lieberman, are you asleep?"

"What! No. Not at all, Mr. McNeal."

"Orion Oil doesn't pay you a $1000 a day to sleep!"

"Oh no! I was just resting my eyes. I usually meditate for a few moments after lunch," I explained. "What time is it?"

"Time to wake up. It's 2:30. Time to write the spec sheet for the new vacuum jet system for the No. 2 vacuum tower."

"Sure, Mr. McNeal. I'll get right on it."

"And get a sample of gas from the seal drum. And don't kill yourself doing it. That cracked gas is loaded with $H_2S$."

"I'll be careful. I'll get Henry to help me."

"Have that flow meter on that cracked gas zeroed-out. Fill out the spec sheet for 50% over the current rate. We don't want the new jet system to be too small for future expanded vacuum tower charge rates."

"Yes sir! I'll follow through. But Mr. McNeal, what's wrong with the existing jets? Maybe the diffuser throat's worn out? Could be wet steam? The steam nozzle could be suffering from erosion? No sense buying a whole new jet system just because the existing jets are...."

"No, Lieberman," Dave McNeal interrupted, "The jet system's working on its performance curve. The Graham Jet rep checked it out. It's just that the system's overloaded with too much cracked gas. Look at the curves yourself."

"Sure, Mr. McNeal. You're right. I've looked at the curves. You're right, sir. The jet system's overloaded. That's why we can't pull a good vacuum. The vacuum tower pressure of 6 psia (see Figure 22-2) is 302 mmHg. That's awfully high. That's why Orion Oil's asphault product is full of light gas oil."

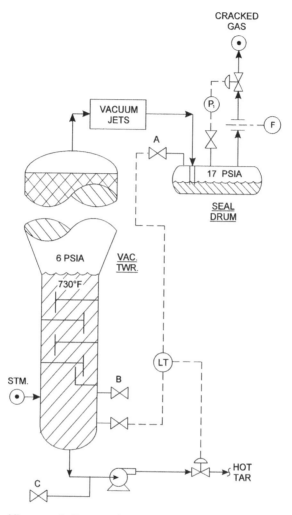

***Figure 22-2*** *Mis connected bottoms level upper tap causes cracked gas flow to double.*

"It's that Venezuelan crude we're running," said Mr. McNeal. "Makes lots of cracked gas. I explained this all to Dr. Beavorbrook, but he told me to buy new jets. That Venezuelan crude's real cheap."

"Yes, sir. Dr. Beavorbrook himself! Well, I'll get right on it. But maybe it's not too much crack gas. Could be we have an air leak?" I asked Mr. McNeal.

"Stop wasting my time!" Dave McNeal screamed. "I checked that cracked gas for nitrogen and oxygen. There ain't none. Just methane, ethane, ethylene, propane, propylene, hydrogen, and lots of hydrogen sulfide. That Venezuelan tar just thermally cracks real easily and makes lots of cracked gas and $H_2S$. Dr. Beavorbrook himself told me that."

## EFFECT OF RESIDENCE TIME ON CRACKED GAS FLOW

"Mr. Henry, I need to get a sample of gas from the seal drum."

"Okay, son. But you be real careful. Best to put on your Scot Air Pak. One breath of that stuff will knock you off your feet."

While I struggled with the straps on the Air Pack face mask, Mr. Henry asked, "Son, you all is an engineer? Well, I got me an engineering question. Why is it that when I raises up the pressure in my seal drum (see Figure 22-2), that the cracked gas flow goes up a lot?"

"Maybe the higher pressure makes the flowmeter read high," I suggested.

"Nope. Ain't so. Higher pressures slows down that there gas velocity through the flowmeter orifice plate. Makes a meter read low, not high. Nope, Mr. Norm. That ain't right."

John Leroy Henry, Jr., had worked at Orion for 40 years. His dad had helped on the startup of the No. 1 pipe still during World War II. Mr. Henry had crushed my fingers with his steely callous grip when we shook hands 10 minutes ago. His boots appeared to have been scavenged from a dumpster. "Look here, Mr. Engineer, let me carry you out an show you."

## MISLOCATED LEVEL TAP

"Mr. Henry, what's that connection on the seal drum for (i.e., connection A, Figure 22-2)?" I asked.

"Well, Mr. Norm. That's there valve A is the connection for the top tap of the level transmitter (LT) from the bottom of the vacuum tower."

"Shouldn't that connection be on valve B, at the bottom of the vacuum tower? Just a bit below the stripping steam inlet?"

"Well, it should be, but it sure ain't. Want to know why?"

"Sure, Mr. Henry, why?" I asked.

"Cause," laughed chief operator John Leroy Henry, Jr., "Lord Beavorbrook done designed it that away when he was the unit engineer out here in 1975."

"Mr. Henry, let me explain why a higher pressure in the seal drum makes more cracked gas:

*Step 1.* Let's assume that the seal drum pressure increases from 16 psia to 17 psia.

*Step 2.* The differential pressure across the level transmitter (LT) is reduced.

*Step 3.* As the differential falls, the measured level drops.

*Step 4.* As the measured level drops, the level control valve closes, to restore the falling measured tower bottom's level.

*Step 5.* The real liquid level in the tower increases, as does the residence time of the hot tar in the bottom of the vacuum tower.

*Step 6.* Thermal cracking of tar is a function of residence time and temperature. At 730°F, increasing the residence time of the tar in the bottom of the vacuum tower increased the production of cracked gas, which overloaded the jet system.

"Mr. Norm, that all makes real good sense. But, I wonder, then, what's the actual level in the bottom of the vacuum tower? I bet that it's real, real high."

"John, screw a vacuum pressure gauge on the bottom's pump suction at connection C. Then we'll see what the suction head pressure is. For every 1 psi of head pressure above the tower pressure, you can figure we have $2\frac{1}{2}$ ft of liquid height above grade."

"Look, Mr. Norm! We done pegged out that there gauge you had me install just now. That liquid level must be backed up above all them stripping trays in the tower," John Henry's voice was shaking with excitement, "You all want me to get that level down some?"

"Yeah. Lower the level down to 10 psia [of course, at sea level, I am talking about reading an ordinary vacuum gauge at about 9 in. Hg vacuum]. That will get the liquid below the trays, but we will still have plenty of NPSH (net positive suction pressure) for the pump. But John, let's go nice and slow," I advised.

"Slow?" John Henry shouted above the roaring jets, "Slow? We ain't going to go slow, Mr. Norm! My great grandpa was a steel-driving man. He died with a hammer in his hand. He done beat that coal drilling steam machine. Happened long time ago. Out in them ole coal fields in western Pennyslvania. Died with a hammer in his hand. But I ain't forgot him!"

John Henry, descendent of an all-but-forgotten American hero, drew the liquid level down rapidly. I later calculated that the liquid hold time in the tower bottoms was reduced from roughly 15 minutes to 3 minutes. The flow of cracked gas dropped from 250,000 scf/day to 110,000 scf/day. As the vacuum improved in the tower due to the unloading of the jet system, the control center panel operator increased the vacuum heater outlet temperature by 20°F. The light gas oil content of the tower bottom's product dropped to the required specification to produce paving-grade asphalt.

## PRODUCTIVE PROCRASTINATION

I explained the situation as best I could to Dave McNeal. I explained about the misplaced top-level tap; about the high liquid level; about the excess thermal cracking; about John Henry and his great grandfather's hammer. But mainly I explained about the reduced flow of cracked gas load to the jets as we reduced the tower bottom's level.

"Look, Lieberman. A spec sheet for the new vacuum jet system is still required. Lord, er I mean Dr. Beavorbrook, expects this project to move forward."

"Okay, Mr. McNeal, you're the boss."

"Yeah! But let's go out for competitive bids for the new vacuum system. You need to get a quotation from this outfit in Ulan Bator."

"Ulan what? Where's that?" I asked.

"In Outer Mongolia. I read about a factory that they're building in the Gobi Desert."

"Really! You've got their email address, Mr. McNeal?"

"No, Lieberman. Just send the spec sheets off by regular mail. And don't waste money on those fancy airmail stamps either. And forget about that level tap business on the seal drum. That's an instrumentation engineering problem, not a process engineering problem. Lieberman, you following any of this or are you falling asleep again?"

## Acknowledgment

The author is indebted to Mr. Garry Carlin, of Murphy Oil, New Orleans, for that part of the story pertaining to the mislocated top-level control tap. Garry is a former student of a troubleshooting class I taught at the Total-Fina Refinery in Milford Haven in Wales, a refinery I helped design, back in 1969, for the Amoco International Oil Company.

# Chapter 23

# Summary Checklist for Reuse of Process Equipment

This chapter summarizes methods I have used in refinery and petrochemical plant process retrofit and expansion projects. The objective is always to reuse the existing equipment rather than to purchase new equipment. As long as a piece of process equipment can be used in place, internal modifications will normally have a minimal environmental impact compared to installation of a new pump, vessel, or heat exchanger. As may be seen from my tabulation below, expanding existing process equipment will also improve energy utilization of fired heaters, compressors, motors, pumps, and turbines. We should also try to avoid installing new plant utility systems, as described below.

## EQUIPMENT CHECKLISTS

### Fired Heaters

1. Minimize the draft to 0.05 to 0.10 in. $H_2O$ as measured below the bottom row of convective tubes. Pinch on the stack damper and open the air registers.
2. Reduce air leaks in the convective section.
3. Blow or wash soot off the finned and studded tubes in the convective tube banks and shock tubes.
4. If burning fuel oil, knock or blow vanadium ash deposits off the radiant tubes.
5. Increase excess air. Unfortunately, this wastes energy.

*Process Engineering for a Small Planet: How to Reuse, Re-Purpose, and Retrofit Existing Process Equipment,* By Norman P. Lieberman
Copyright © 2010 John Wiley & Sons, Inc.

*Note*: Air preheaters will save energy but will reduce the overall heater capacity if the heater is limited by the radiant heat density or the radiant tube coking rate.

### Heat Exchangers

1. Re-tube the bundle from carbon steel to stainless steel so as to retard fouling deposit accumulation.
2. Increase the tube-side passes if the pressure drop permits.
3. Replace the tube bundle with twisted tubes or helical baffle tube bundles.
4. Thermally shock (i.e., spall) tubes so as to remove the fouling deposits.
5. Back-flush or acid clean the tube side of water coolers.
6. Replace the bundles with low-fin tubes if both the shell and tube sides are clean.
7. For reboilers with the process fluid on the shell side, lightly sand-blast the tubes' exterior, to provide nucleation sites for bubble formation.
8. Add shell-side seal strips for improved shell-side sensible heat transfer.

### Air Coolers

1. Maximize the fan blade pitch.
2. Increase the size of the motor pulley wheel. Check the torque rating of the fan blades first, and the motor FLA point.
3. Wash the fins from underneath the fin tube bundle.
4. Use vane tip seals to reduce air recirculation through the forced-draft fan.
5. Seal leaks between the tube bundle itself and the tube bundle support structure.
6. Reverse the rotation of the fan for a few minutes, to blow loose deposits off the underside of the tube fins.

### Distillation Tower Trays

1. Replace valve or sieve tray decks with Sulzer push-type MVG tray panels. Reuse existing downcomers and bolting bars.
2. Convert existing side downcomers on two-pass trays to segmental downcomers (but not segmental weirs).
3. Reuse the downpour areas, as bubble areas, using Nye-type inserts, or panels, below the downcomers.
4. Increase the downcomer clearances, but do not unseal the downcomers.
5. Increase the hole area by drilling holes in the tray deck periphery, but do not exceed 14% of the tray bubble areas.
6. Increase the downcomer areas by sloping downcomers using Z-bar clamps on existing downcomer bolting bars.

## Vapor–Liquid Separators

1. Retrofit with demisters, but only for clean, noncorrosive services.
2. Inject silicon defoaming agents, but only if the downstream silicon contamination can be tolerated on the catalyst.
3. Improve liquid-level control using radiation-type foam density-level detection, such as neutron backscatter.
4. Control the vessel bottom's pump suction pressure, and drop the liquid level out of the vessel altogether.
5. Raise the vessel operating pressure so as to reduce the vapor velocity.
6. Add a vapor horn or impingement plate at the feed nozzle to dissipate the excessive inlet momentum.
7. Add a chimney tray to promote uniform vertical vapor velocities above the vapor inlet nozzle.

## Centrifugal Pumps

1. Replace the impeller wear ring.
2. Reduce impeller-to-case internal clearances back to the original design specifications.
3. Increase the impeller size. Note that power requirements increase with the cube of the impeller diameter.
4. Increase the trim size in the downstream control valve.
5. Increase the density of fluid pumped by reducing liquid temperature.
6. Subcool liquid at the pump suction, to increase the available net positive suction head.

## Reactors (Fixed-Bed Refinery Hydrotreaters)

1. Reduce upstream use of silicon defoaming agents.
2. Reduce upstream corrosion to reduce catalyst bed plugging.
3. Install baskets on top of the catalyst bed. Immerse the baskets in the top few feet of the catalyst.
4. Avoid exposing olefinic-type feeds to air, which will cause them to form gums and plug the catalyst beds.
5. Minimize entrained asphaultines in the feed, as they cause nickel and vanadium ions to fill the catalyst pores. (*Note:* High-conradson carbon feed that is not due to entrained asphaultines is never bad for the catalyst.)
6. Increase hydrogen partial pressures.
7. Improve mixing between the hydrogen and interbed flows. (See U.S. patent 3,855,068 "Apparatus for Inter-bed Mixing of a Fluid Quench Medium for a Vapor–Liquid Mixture," by N. P. Lieberman, Dec. 17, 1974.)

### Electric Motors

1. Increase the full-limit amp set trip point. (*Caution*: This may reduce the time between motor rewindings.) Only small increases (i.e., less than 10%) are typically made.
2. Clean the screen in back of the motor.
3. Lubricate the motor bearings.
4. Trim the size of the impeller of the pump being driven by the motor.
5. Have the motor coils rewound.

### Gas-Fired Turbines

1. Clean the suction filter combustion air intake screens.
2. Detergent-wash the front-end axial air compressor blades.

### Steam Turbines

1. Make sure that all hand valves are open.
2. Increase the size of the steam inlet nozzles in the nozzle block.
3. Condensate-wash or degrease the turbine wheels.
4. Make sure that the governor speed control valve is 100% open, by minimizing the $\Delta P$ value across the governor.
5. For condensing steam turbines, improve the vacuum in the surface condenser.

### Centrifugal Compressors

1. Increase the molecular weight of the gas if limited by the maximum rated compressor speed.
2. Minimize the molecular weight of the gas if limited by the compressor's driver horsepower.
3. Spray a heavier liquid into the compressor suction to keep the wheels from drying out and fouling.
4. Suction-throttle the gas if limited by the compressor's driver horsepower.
5. Clean the rotor wheels, especially if limited by the compressor's maximum rated speed.

### Reciprocating Compressors

1. Remove the pulsation damper plates (or reduce their $\Delta P$ value by bigger orifices) from the suction and discharge lines.
2. Obtain an indicator card survey to identify bad valves and leaking piston rings.
3. Install adjustable head-end uploaders and remove valve disabler unloaders.
4. Optimize the spring tension in valves to minimize leakage, but without excessive pulsation losses.

### Air Blowers

1. Bypass the air-intake silencer if the measured pressure drop is more than a few centimeters of water.
2. Clean the air-intake suction filters and modify the intake screen to reduce $\Delta P$.
3. Water-wash the rotor wheels.
4. Control the flow on the air intake suction, if limited by motor amps, rather than on the discharge airflow control valve.

### Water–Oil Separators

1. Increase the height of the riser inside the separator vessel.
2. Introduce the two-phase mixture via a distributor, which reduces turbulence in the separator vessel.
3. Back-flush the water draw-off connection with clean water.
4. Back-flush the interface-level tap connections with clean water to maintain an accurate level indication between the aqueous and hydrocarbon phases.
5. For electrical precipitators, clean the insulators so as to maximize the voltage across the grid, without excessive amperage.

### Piping Pressure Losses

1. Remember that piping pressure losses increase inversely with the diameter raised to the fifth power. Example: One mile of 3-in. pipe has the same $\Delta P$ value as 32 miles of 6-in. pipe.
2. Repair concrete liners in seawater piping systems used for cooling.
3. Chemically clean freshwater circulating cooling piping systems.
4. Replace the none full-ported isolation valves. An 80% ported valve has $2\frac{1}{2}$ times the $\Delta P$, as does a full ported-valve.

## EQUIPMENT LIMITATIONS

### Mechanical Limitations

1. Pump temperature limits are typically due to mechanical seal design temperature limitations. Change the seal, not the pump.
2. Pressure limitations can often be overcome by rerating MAWP using corrosion allowance to recalculate the mechanical strength of the vessel.
3. Specified heater temperature limitations are often process specifications rather than actual mechanical temperature limitation.

### Utility System Limitations

1. *Instrument air.* Find and fix air leaks. Disconnect air-operated tools and other noninstrument users.

2. *Cooling water*. Repair holes in cooling tower distribution grids. Clean and repair cooling cell distribution decks. Repair slipping belts on fans. Isolate water coolers not in service.

3. *Steam systems*. Repair weatherproofing and insulation. Check the increase in the plant steam demand during a rainstorm.

4. *Electric power*. Eliminate control valves and replace with variable-speed, frequency-adjusted electric motors. Trim impeller sizes on pumps. Avoid compressor spill-back control, and use suction throttling.

5. *Process water*. Stripped sour water can be used as boiler feedwater for generating low-pressure steam. Steam condensate recovery should be 70 to 80%, not 20 to 30%! Check cooling-water cycles of concentration to minimize the cooling tower blowdown rates.

## SAFETY NOTE

Sometimes, my objective in the reuse of process equipment requires inspecting tower internals to determine plugged and coked internals. The following recent story is a reminder to be careful when performing this work.

Kumar, the unit engineer, and I were inspecting a wash oil spray header for plugged nozzles and proper spray pattern in the gas oil section of a large coker fractionator (Figure 23-1). Based on coking of the wash oil grid, we expected to find several of the nozzles plugged. Our plan was to crawl under the spray header and observe the water flow of the spray nozzles. To access the wash oil header, we crawled down through a chimney. We were outfitted with oversized slicker suits, rubber boots, safety glasses, and hard hats.

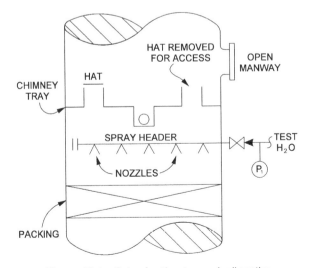

***Figure 23-1*** Coker fractionator wash oil section.

The fractionator had been open for three days. Water and cleaning solutions had been circulated through the spray header, and entry permits were in order. Kumar, being 40 years my junior, slipped through the chimney tray easily. The combination of old age and the slicker suit made it a challenge for me to follow him through the chimney. Regardless, after 10 minutes of effort, I was sitting on the packed bed with my head poking through the spray header. I had committed my first unsafe act. I had entered a confined space that I could not easily exit. I suspected that I could not climb back up the chimney without Kumar's help.

Now for the almost fatal error. As we suspected, some of the nozzles were plugged. I had asked a pipefitter to loosen the spray nozzles so that I could remove the plugged nozzles without tools. But those sections of the spray header piping upstream of the plugged nozzles had never been cleared of hydrocarbons with flush water.

The next error was one of communication. I wanted the water pressure at the spray header inlet to be 10 psig, to match the normal spray nozzle $\Delta P$ of 10 psi. But this information was not relayed to the responsible operator. Thus, we had the full water system pressure of 100 psig at the spray header inlet. As a result of the 100-psi pressure drop through the nozzles:

- The nozzles made a roaring sound that prevented Kumar, who was on the other side of the tower, from hearing me.
- The nozzle produced a mist that fogged my safety glasses. Even after removing them, I could see only 2 or 3 ft.
- The high water pressure blew hydrocarbons, trapped in the spray header lateral arms, out of threads around the loosened spray nozzles.

Kumar's $H_2S$ alarm went off at once. I, as a contractor, did not have an $H_2S$ alarm. Kumar decided to wait a few minutes to see if the $H_2S$ alarm cleared. Not a good decision on his part! Within 2 minutes, the hydrocarbon fumes had made breathing difficult. My eyes were burning, I was coughing, and I became confused. In retrospect, both Kumar and I had suffered a loss of judgment. I recall thinking, "I can't get out myself, and Kumar can't hear or see me."

I could still communicate with the "hole watch" man. His $H_2S$ alarm had also gone off outside the manway. I asked him to shut off the water and help pull me through the chimney. Stopping the water flow did not reduce the overwhelming hydrocarbon odor but did allow me to tell Kumar that we had to get out of the tower immediately. Unfortunately, Kumar could not do so, because I was stuck in the chimney. I was close to the manway and could get some fresh air. Trapped below me, Kumar was breathing hydrocarbon vapors.

If we had more foresight, we could have provided an alternate exit by removing an additional hat above one of the other chimneys.

With the hole watch guy pulling and Kumar pushing, I was extracted from the chimney. Both Kumar and I had headaches and felt tired the next day. The safety lessons I learned from this "near miss" are:

1. Do not enter a confined space if it is very difficult to obtain access. If the unexpected happens, getting out will be just as difficult and take just as long.

2. Do not wear protective clothing that inhibits movement or that catches on metal parts. For example, my rubber boots filled with water, and I wasn't able to pull my way through the chimney, due to their additional weight.

3. If a screwed or flanged connection is broken inside a vessel, it should be water-washed again and new entry permits issued. Breaking any process piping connections inside a vessel should void existing hot work or entry permits.

4. Contract personnel should have the same $H_2S$ monitors as those issued to refinery personnel, even if the contractor will be working closely with the refinery representative.

5. When an $H_2S$ alarm goes off, the person should not try to exercise judgment as to whether evacuation is really necessary. Just clear out!

6. Alternate escape routes should be established.

7. If a parameter is to be adjusted (i.e., the spray water pressure), make sure that the operator actually controlling the parameter is advised as to his or her responsibilities.

## CHECKLIST SUMMARY

As the reader may well imagine, after 46 years I have developed many hundreds of such methods as described above, to reuse existing process equipment in expanded service. These methods are, to a large extent, detailed in my books, which are listed in the preface. The general idea that I apply has always been the same. Use the existing equipment in place. Reusing process equipment in a new location is almost as wasteful as building new process equipment, as new piping, valving, instrumentation, and foundations are still going to be required. Whenever I start a new project, I always say to myself, "What I have now is what I'll use." Only as a last resort will I take my red pencil and sketch in on the process flowsheet a new heat exchanger, vessel, turbine, or centrifugal pump.

# Chapter 24

# *Water–Hydrocarbon Separation: Corrosive Effects of Water*

"Mister, do you like to sit next to the window?"

"What? Yes, I guess so."

"Me, too," she said. "I love to fly on airplanes! Don't you?"

"Want to change seats?" I asked.

"Okay, but get my crayons and coloring book down first. How old are you?" she asked.

"Me? I'm 67."

"I'm almost 7. My birthday was last week. I'm Becky. My mom and my brothers and Aunt Lilly are meeting me in Dallas. I'm thirsty. Are you really 67? That's super-old. Are you gonna die soon? Can you get me juice? I like apple. I'm real thirsty."

Becky sipped sparingly on her apple juice. "Mister, I'm hungry."

"Becky, they're not serving lunch yet."

"Tell them a child is hungry," she whimpered softly at the point of tears.

Becky nibbled carefully on a carrot stick and wrapped the sandwich in her napkin. "I'm saving the sandwich for my mom. She's meeting me at the airport. Are we almost there?"

I had fallen asleep.

"Hey, mister." Becky shook my arm to make sure that I was properly awake. "You really old people are also really smart. You're really smart, too?"

"I guess so. Becky, where's your lunch?"

*Process Engineering for a Small Planet: How to Reuse, Re-Purpose, and Retrofit Existing Process Equipment.* By Norman P. Lieberman
Copyright © 2010 John Wiley & Sons, Inc.

"Oh! I packed it up with my crayons. I'm too full. But mister, I'll ask you a question if you're so smart."

"I thought you were real hungry."

"I'm too full. But I heard my dad talking about greenhouses and gases and globals and glaciers. Did you know, mister, that Greenland is melting? If the North Pole melts, where will Santa Claus live? Did you know Santa has a giant factory to make toys in the North Pole? Thousands and millions of elves do the work. But if the North Pole melts, the elves will drown in the ocean. I'm getting a doll with blue eyes for Christmas. But what's $CO_2$ anyway? I think $CO_2$ is really bad—like getting the stomach flu. I got the flu, too."

"Well Becky, $CO_2$ is. . . ."

"Mister, if you're so smart like you just said you are, can't you figure out how to get all that $CO_2$ and put it in a bottle and rocket it away to Mars?"

"Well, Becky, it's not that simple to. . . ."

"Mister, tell me a story. I'm tired. Tell me a story with a happy ending. About how you're going to get rid of all those mean $CO_2$'s. You're so old that you've got to be really smart. Do you know any stories?"

## BECKY'S STORY

"Once upon a time in a faraway land, there grew a giant refinery with beautiful, tall distillation towers stretching high into the sky. One day a wicked witch decided to destroy the beautiful distillation towers using hydrogen-assisted stress corrosion cracking. To aid her in this evil scheme, she brought her equally evil companions:

- Hydrogen cyanide, HCN
- Hydrogen sulfide, $H_2S$
- Hydrochloric acid, HCl

"Mister, where did her evil companions come from?"

"Cracking reactions, Becky. From the salt, nitrogen, and sulfur in the feed to cokers, fluid cat crackers, and visbreakers."

"I like cats. I'm especially sure I would like a fluid cat. Gloria has a gray cat."

"So the evil witch caused the vessel walls to crack and also to blister. Hydrogen blistering and hydrogen cracking are caused by the same chemical mechanism. You see Becky, when steel corrodes, there are always two products of corrosion: the salt and a hydrogen ion, or a proton. For example:

$$H_2S + Fe = Fe(HS)^- + H^+$$

The $Fe(HS)^-$ ion turns into iron sulfide, a black slippery solid. The $H^+$ ion is invisible and turns into molecular hydrogen.

"I have an invisible friend, too. Her name is Lulu."

Becky was not too good at listening. Anyway, I continued, "ordinarily the ionic hydrogen combines at the vessel wall to form molecular hydrogen, which then leaves with the product. But in the presence of hydrogen cyanide, this does not happen. What does happen is that the proton dissolves in the steel wall of the vessel, in the same sense that salt dissolves in water. This is not usually a problem. The proton simply passes to the outside of the vessel. There it will combine with another friendly proton to form molecular hydrogen, which floats safely away."

"But suppose the proton gets lost in the steel," Becky asked. "I got lost once in the mall, but I didn't cry."

"That does happen. At imperfections in the lattice structure of the steel, but especially at the heat-affected zones of welds that have not been postweld heat treated or stress relieved. Then one or all of several bad things will happen to the vessel wall.

- *Hydrogen blistering.* This looks like an old piece of plywood that has been soaked for months. The steel becomes delaminated.
- *Microcracking.* Small cracks not visible to the naked eye permeate the vessel wall and cause a loss in structural strength.
- *Hydrogen-assisted stress corrosion cracking.* A crack can rapidly follow the heat-affected zone of a weld and cause rapid and total vessel failure. Seventeen people were killed at the Unocal Refinery in Chicago by such a failure."

"Mister, why does the evil witch want to hurt the vessels? Is it because they are too tall? My brothers are giants."

"No Becky. The witch doesn't like people. She knows that if she ruins the vessel, the refinery will have to buy a new one. Then iron ore will have to be mined and coal will have to be dug up to make the steel. Then the steel will have to be melted to make the plate for the new vessel. All of this generates $CO_2$ and other greenhouse gases, which go off into the air. The $CO_2$ gets the air hotter, which then melts Greenland."

"I think someone should stop that witch. Maybe killing her would be best. We could drop a house on her."

## COMBATING CORROSION

"Well, Becky, the best way to stop corrosion is to eliminate an aqueous environment—I mean, get rid of all the water. Gasoline, when it's cool and dry, isn't corrosive. But water, when it forms a separate phase, extracts HCN, HCl, ammonia, and $H_2S$ from the cracked gases.

The HCl makes $FeCl_2$, which then reacts with the $H_2S$ to make $Fe(HS)_2$ and protons and new HCl. The HCN kind of pushes the protons into the steel. The protons join up and make molecular hydrogen inside the vessel wall at imperfections in the lattice structure. The hydrogen creates gas pressure, which causes the cracks and blisters.

So the witch decided to destroy the beautiful, tall distillation towers by causing water to get trapped inside the towers. The evil witch did this is three ways:

1. Plugging off the side draw-off on a water accumulation tray.
2. Plugging off the water draw-off boot on the overhead reflux drum.
3. Reducing the separation efficiency between the water and hydrocarbons in the distillation tower feed drums."

"Mister, is this the good witch from the North, Glendela, or the evil witch from the East?"

"Becky, it's the East witch, Moriah, and I'll tell you what she did."

## HYDROCARBON–WATER SEPARATION EFFICIENCY

"Give me your coloring book and crayons and I'll draw you a picture (Figure 24-1). That's the feed drum to the Saturn coker naphtha stripper in Los Angeles, California. The drum is supposed to settle-out free water from the liquid hydrocarbons before the hydrocarbons are pumped to the stripper tower. But it couldn't do this efficiently because the riser in the drum was too short."

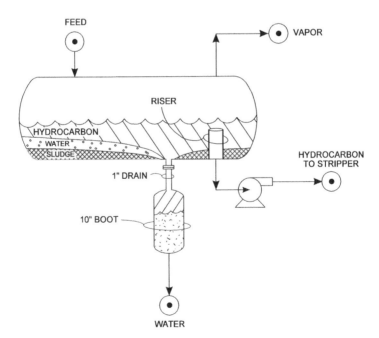

*Figure 24-1* *Water–hydrocarbon gravity settler.*

"What's the riser for? I'm kind of short too. You didn't notice cause I'm sitting up very straight."

"The riser permits the water to run into the boot but helps keep the hydrocarbons liquid out of the boot. Also, and even more important, it keeps water out of the feed to the stripper. The riser should be at least 12 in. high. Otherwise, as iron sulfide sludge accumulates in the bottom of the drum, the water in the feed will not have time to separate from the hydrocarbon liquid. Figure 24-1 also shows a common design error. The 10-in. boot is connected to the drum with a 1-in. drain line. This makes it possible to locate the boot at grade. As a former plant operator, having the water draw-off boot at grade was a nice convenience, as I didn't have to climb up to check it. However, at the Los Angeles stripper feed drum, I found that this was not a good idea. The small 1-in. connection plugged with sludge and caused water to be carried into the top of the riser. I fixed this problem by installing a blowback connection so that I could clear the 1-in. drain with high-pressure water connected to the boot.

Sometimes these settlers are simply too small. For design I use a water vertical settling rate of 0.5 ft/min for the droplets of water settling through a low-viscosity (less than 5 cP) hydrocarbon liquid of 0.6 to 0.8 S.G. In practice, this velocity can be a maximum of 1.3 to 1.4 ft/min. Beyond such velocities, aqueous phase carryover can be anticipated."

"Mister, I'm thirsty. Tell that nice lady I want another juice. I like apple juice."

## AVOIDING WATER TRAPS IN STRIPPERS

"Becky, you didn't drink the last juice."

"I want a new juice. Mister, tell me the rest of your story. I like stories with happy endings."

"Here's the problem. Even if the water-settling drum removes all of the free water, some water will still be dissolved in the stripper feed. For 110°F coker naphtha with some aromatics and lots of olefins, the solubility of the water is about 400 ppm by weight."

"Dolphins are like olefins, you know. I love dolphins. They save drowning children and play games. When I grow up, I want to be a dolphin. Do olefins have fins?"

"No, Becky. These are hydrocarbon molecules with a double bond. But the point is that the dissolved water in the feed gets trapped inside the stripper (Figure 24-2). The water is supposed to be drawn off of tray 4 into the boot, but some naphtha (hydrocarbon liquid) is drawn off with the water. The boot interface level control valve only allows water to escape out of the bottom of the boot. The naphtha has to back up to the stripper through the feed line. But if the downward vertical velocity of the water exceeds 1.4 ft/min, the naphtha can't flow countercurrently back to the stripper through the 1-in. line shown in Figure 24-2. Then the naphtha gets trapped inside the boot and prevents water from flowing from tray 4 into the boot."

"But mister, why does the water get trapped inside the stripper? Does Moriah, the wicked witch of the East, lock it up?"

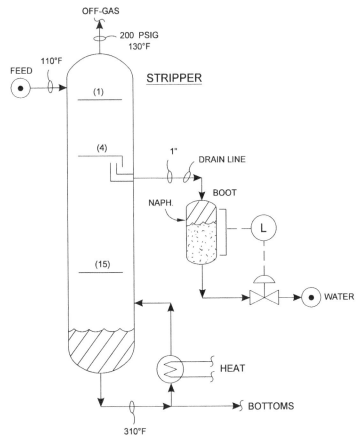

**Figure 24-2** *This 1-in. drain line to water draw off boot is too small to permit naphtha to flow back into the stripper.*

"Well, Becky, the 130°F stripper top temperature is too cold to allow much water vapor to escape overhead. The stripper bottoms temperature of 310°F is too hot to allow the water to escape very easily out the bottoms. So not only does the water get trapped inside the tower, but it can also start to circulate in the reboiler. The water is very corrosive because of the $H_2S$ and attacks the carbon steel reboiler tubes. The water will eat up such tubes in just eight months. That happened at the Los Angeles coker in California."

"The witch's evil helper $H_2S$ ate the tubes for lunch. Does carbon steel taste much like chocolate chips? I dearly love chocolate chips," Becky said as she sipped a few droplets of apple juice.

"To correct the problem of not being able to draw-off the water from tray 4, I changed the boot level control valve to a flow control valve. Then I drew off enough liquid from tray 4, to make sure that the water in the draw-off sump was all drawn off.

Of course, this caused a lot of naphtha to be drawn off with the water. So I installed a new 1-in. line to recycle the liquid flow from the boot back to the feed drum in Figure 24-1. The operators adjusted this flow so that the water drawn off the overhead of the downstream stabilizer tower reflux drum was almost nothing.

This stopped the corrosion in the reboiler, in addition to stopping the hydrogen penetration of the stripper vessel wall. The corrosion in the reboiler was monitored by a corrosion coupon inserted at the bottom of the shell. The hydrogen penetration rate of the vessel wall was monitored by the use of hydrogen probes attached to the exterior of the vessel wall.

## USE OF DUMP VALVES TO PREVENT DRAIN VALVE PLUGGING

The stripper bottoms flowed to a downstream stabilizer tower. The overhead reflux drum also had a water draw-off boot. Sometimes, when the side water draw-off on the stripper was not working properly, water would appear in this stabilizer overhead drum boot. This is okay, provided that the drain valve on the water draw-off boot (Figure 24-3), did not plug-off. The problem was that this drain valve was almost always closed. And while it was closed, the evil witch would cause a small amount of iron sulfide sludge to plug-off the drain valve. If the drain valve was a full-ported valve, Becky, this wouldn't have mattered. But the drain valve was sized to control a flow of only 0.1 gpm. So the port size, or internal trim, was very little. The witch knew this, so she plugged the valve with sludge when it was closed. Then, when it opened, no water could drain out.

The boot would fill up and water would overflow into the riser. Next, the water would be refluxed down the tower and corrode the reboiler, the trays, and the distillation tower itself. And then the. . . ."

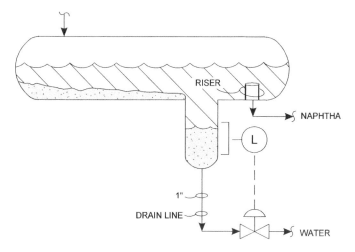

*Figure 24-3*  Low water flow caused drain valve to plug.

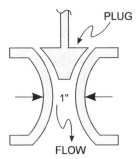

**Figure 24-4**   *Dump valve cannot plug because of large port size.*

"And then bad $CO_2$ would be made," said little Becky, "And escape into the air because we had to make the new steel stuff because Mister Corrosion ate up the old steel stuff!"

"That's right, Becky. And to stop the evil witch's plan, I installed the dump valve shown in Figure 24-4. The valve has a large port, too large for the witch to plug-off. Most of the time the valve is shut. But when the level in the boot pulls the plug up, water rushes out. The level in the boot drops by several feet in a few seconds. The plug drops back into its seat, and the water level slowly creeps back up for hours or even days."

Having modified the stabilizer and the feed drum and the stripper-side water draw-off boot drains, both the stripper and the stabilizer were now operated in a dry and happy environment. The corrosion and hydrogen activity stopped in the reboilers and the tower's steel walls. The evil witch's plans were ruined. The beautiful tall distillation towers were saved and they lived happily ever after in Los Angeles, California. And that's the end of the story."

## CURRENT CO$_2$ LEVELS

"Oh! That was a wonderful story." Becky clapped her small freckled hands in delight. "I love stories with happy endings. Now when I grow up all that bad $CO_2$ in the air is going to be gone. Now Greenland won't melt and polar bears and Santa Claus won't drown in the ocean. I was so worried about Santa's elves, you know. I'm learning how to swim, but Santa can't swim. He's too fat."

"Well Becky. That's just one story. The other stories do not have such a happy ending. Here's the problem:

- Global fossil carbon emissions are increasing more rapidly every year, at a rate of $3\frac{1}{2}\%$ in 2009. From the time I was born in 1942 until today, carbon emissions due to human activities have increased from 1000 to 8000 million metric tons of carbon per year.

- The production, refining, and distribution of coal, natural gas, gasoline, jet fuel, diesel oil, and other fossil fuels consume about $8\frac{1}{2}$% of fossil fuel production themselves.
- Currently, $CO_2$ emissions from fossil fuels are divided up among:
  - Coal: 35%
  - Crude oil: 40%
  - Natural gas: 20%
  - Cement: 3%

Crude oil $CO_2$ emissions are increasing only 1.0 to $1\frac{1}{2}$% per year. Natural gas consumption is going up a little faster, at 2.0 to 3.0% per year. But coal consumption is surging ahead. In remote provinces of China, India, and other power-hungry countries, conventional coal-fired power plants are being built really fast, at a rate of one per week. Even though $CO_2$ emissions are pretty constant in the United States and Europe, carbon emissions from coal and other fossil fuels in China, India, Brazil, and Indonesia are causing our little planet to drown in $CO_2$."

"Clean coal!" Becky shouted, "It's on TV. Mister, don't you watch TV?"

"Becky, clean coal is a real fairy tale. It won't happen. The oil, coal, and natural gas companies are lying to us. Here's the truth. $CO_2$ in the atmosphere is now 390 ppm (383.7 in 2007, 385.5 in 2008, and 387.7 in August 2009) and increasing at a rate of 0.52% per year (2.0 to $2\frac{1}{2}$ ppm by volume) every year. The average surface temperature of the world's seas was 1.06°F hotter in 2009 than the average for the twentieth century [2]. If this trend continues, as now appears very likely, the $CO_2$ in the air by the time you are my age will be 640 ppm. Then the average temperature on the Earth will be about 5 to 6°F hotter than it is today. Then Greenland, the Arctic, and the Antarctic ice shelves will mainly be melted, and my home in New Orleans will be under 30 feet of water."

"But, mister, you're so old. I'll never get as old as you. You must be even older than Grandma. Anyway, I hate putting on my snow suit. The zippers always get stuck. You know, my sled is broken, so I don't care about winter or not having snow anymore."

The flight attendant came to collect Becky's apple juice and we landed in Dallas. Becky and I were the first passengers to disembark. As we left our seats, little Becky turned to look at me. She stared deeply into my eyes and fear spread across her beautiful face.

"A monster!," she screamed and ran up the jetway passage. When I emerged into the terminal after her, she was surrounded by her family. Her three brothers fought to hug her first.

"Little sister, we missed you."

An older lady cried with joy, "Becky, welcome home. I brought you a new doll," while she pushed a younger woman aside to hug Becky.

Becky just smiled and said to the younger, tired looking, but well-dressed lady, "Mom, I'm thirsty! I need an apple juice right away! Can you buy me a fluid pussycat? There was a terrible monster on the plane. Can we go right to the park? I want to play in the sunshine on the swings with Gloria before it gets too hot. Greenland is melting today, you know. The monster told me that a witch is making a greenhouse all around the world to drown Santa's elves. I hate monsters and $CO_2$'s and witches. Can we get apple juice that's in a blue bottle with a red cap? I'm so thirsty. Look mom, my new doll has blue eyes."

## A FINAL OBSERVATION

To a large extent we are locked into a minimum degree of environmental degradation. Realistically, putting aside the remote possibilities, which are for the distant future, what can Becky expect when she is 36 years old and has her own 6-year-old child?

- The ambient $CO_2$ concentration, which has increased by 100 ppm in the last 120 years, will increase by another 100 ppm in the next 30 years.
- Land surface temperatures, which have increased about 1°F in the last 100 years, will increase by perhaps another 1°F. The land temperature increase relative to the increase in $CO_2$ will accelerate, due to warming ocean surface temperature and reduced solar reflection from atmospheric sulfates.
- Ocean levels will rise by 6 in., due to thermal expansion, plus 12 in. due to glacial melting, for a total of 18 in.

Improvements in electrical generation by wind, solar, and nuclear energy are being more than offset by new coal-fired power plant construction. New fuel-efficient vehicles are more than offset by the growing number of vehicles. Improvements in the use of nitrogen-based fertilizers are more than offset by the growing use of fertilizers to feed an expanding human population.

The escalating values for atmospheric $CO_2$, ambient temperature, and sea levels I have quoted above may get far worse. As crude oil production levels out at 90,000,000 bsd, I fear that shale oil production, coal-to-liquids, tar sands, conversion to synthetic crude, and natural gas-to-liquids projects will become increasingly prevalent. The Germans produced half their aviation gasoline in World War II by synthetic means. I visited these partially intact units in 1984: an older coal-to-distillates project.

I have some actual data from one of the world's largest coal-to-liquids facilities (Sasol in South Africa). I taught a seminar in South Africa in 2009 and a couple of their experienced engineers calculated the following for me:

- *Input from coal:* – 100 Btu
- *Output in the form of usable products* (liquid fuels, chemicals, anode-grade coke, intermediates, etc.): 32 to 35 Btu

That's a pretty awful carbon footprint compared to crude oil and natural gas. But coal reserves are very large, and crude and gas reserves are quite limited. The CVR Energy plant in Kansas converts 100 Btu of coke to about 50 Btu of hydrogen. This seemed really good to me until the tech manager added that the refinery also supplied all the utilities to the conversion plant.

As I write these words, I'm flying to Ft. McMurray to work on an expansion of a tar sands project. If the major oil companies also restart their Rio Blanco shale oil production project in Fruita, Colorado, you can ignore my projections for the environmental future of our small planet. I explained my concerns to my youngest daughter.

"Dad," Irene said, "That's not much of an advertisement for having kids."

No, Irene, it's not. I had tried to explain this to Becky. But like the senior management of many hydrocarbon processing industries, she became angry and decided that I was a monster.

May the Creator of the Universe protect our innocent children from the folly of humankind.

## REFERENCES

1. Wikipedia. *Greenhouse Gas: Natural and Anthropogenic Sources,* Apr. 2008.
2. National Oceanic and Atmospheric Administration Climate Data Center.

*Appendix*

# Solar Power Potential

I gathered the following data from one of my students (an operator at the Total-Elf refinery in Holland):

- *Unit size:* twenty 11-ft$^2$ solar panels
- *Rated capacity:* 3000 W
- *Location:* Holland
- *Cost* (as installed by contractor, including battery storage, inverter, and insurance): 20,000 euros ($28,000)
- *Installed:* May 2007
- *Total metered electricity generated after one year of operation:* 3540 kWh
- *Average hourly power generation:* 405 W

Note that the actual generation rate is only $13\frac{1}{2}\%$ of the rated capacity, due to darkness, clouds, the angle of the sun, and the panels being less than 100% clean. Let's assume that this is a project with a 10-year capital recapture. This means that the cost of the power will be about $1.30 per kilowatt. The maintenance costs are covered by the insurance policy, which was included in the installation.

In comparison, I am paying 10 cents/kW in New Orleans for power generated from natural gas and nuclear energy. Of course, this comparison excludes the effects of government subsidies, which is why my friend from Holland installed his solar panels. Also, the figure of 10 cents reflects natural gas costs of $6.00 per million Btu.

I have purchased and installed several solar panels myself as an experiment.

- *Rated capacity:* 120 W (total)
- *Location:* New Orleans, Louisiana
- *Cost:* $1000 for panels, inverter, and battery storage
- *Unit size:* eight 1 ft × 3 ft panels

On a bright day, if I move the panels to face directly into the sun, they do generate 120 W of power at 12 V. However, by the time I convert this from the $12\frac{1}{2}$ V storage battery to 115 V and don't bother to reposition my panels for the sun's movement in the sky, I have found over a two-week period in September that I only generated about 20 W of usable power on average.

So, renewable energy is practical, but without taking into account environmental effects, it seems very expensive to me. But it's this environmental effect that's probably what I should include in my economic analysis of solar power. You all know: like the cost of scrubbing the $CO_2$ out of the flue gas from the gas-fired power station down by the west bank of the Mississippi River in St. Charles Parish.

# *Index*

Printed and bound by CPI Group (UK) Ltd, Croydon, CR0 4YY

27/10/2024

14580260-0003